THE DANUBE EMPIRE

INTERSECTIONS:
HISTORIES OF ENVIRONMENT, SCIENCE, AND TECHNOLOGY IN THE ANTHROPOCENE

Sarah Elkind and Finn Arne Jørgensen, Editors

THE DANUBE EMPIRE

An Environmental History
of Habsburg State-Building and Civic Engagement

ROBERT SHIELDS MEVISSEN

University of Pittsburgh Press

Published by the University of Pittsburgh Press, Pittsburgh, Pa., 15260
Copyright © 2025, University of Pittsburgh Press
All rights reserved

Printed on acid-free paper
10 9 8 7 6 5 4 3 2 1

Cataloging-in-Publication data is available from the Library of Congress

Hardcover: 978-0-8229-4874-2
Paperback: 978-0-8229-6779-8

Cover art: Unknown photographer, *Anlegestelle an der Unteren Donaulände*, 1903. Nordico Stadtmuseum, Linz, Austria. Archival number: NA-051009.
Cover design: Melissa Dias-Mandoly

Publisher: University of Pittsburgh Press, 7500 Thomas Blvd., 4th floor, Pittsburgh, PA 15260, United States, www.upittpress.org
EU Authorized Representative: Easy Access System Europe, Mustamäe tee 50, 10621 Tallinn, Estonia, gpsr.requests@easproject.com

Meiner wundervollen Frau, Elisabeth,
und unseren weltbesten Kindern gewidmet

CONTENTS

Acknowledgments

ix

INTRODUCTION

The Danube Empire

3

CHAPTER 1

Creating the Imperial Danube

23

CHAPTER 2

The Danube as Life Artery

57

CHAPTER 3

The Danube as a People Network

89

CHAPTER 4
Overcoming Danubian Dangers
120

CHAPTER 5
Act Locally, Think Imperially
149

CONCLUSION
Collective Action and the Common Good
184

Notes
189

Selected Bibliography
225

Index
255

ACKNOWLEDGMENTS

Gratitude is a bit like slices of cake or cups of tea; one hopes to do an adequate job of finding moments to share it with the right people over the years. First and foremost, I am immensely grateful to Jim Shedel, whose engaging conversations, probing questions, and generous suggestions were the proverbial iron sharpening my thought process. His insistence that I learn Hungarian has led to many grammatical and literal adventures, and his ongoing support has been instrumental from research to writing to revising. Likewise, had it not been for John McNeill's casual conversations on the basketball court, his introduction to environmental themes and to Austrian scholars, and his unflagging support, I would likely have written a much less interesting book. I want to thank the Austrian Fulbright Commission (and Lonnie Johnson) for the opportunity to spend time in Vienna soaking up history and inspiration for this work. I want to thank the Hungarian Fulbright Commission (and Károly Jókay) for supporting my research and introducing me to many great cultural and historical sites around Hungary. The Department of Education's Foreign Language and Area Studies scholarship allowed me to spend a glorious summer split between Pittsburgh and Debrecen grappling with the intricacies of *a magyar nyelv* and eating *pogácsa*. Verena Winiwarter and Martin Schmid were generous mentors when I was in Vienna; they invited me to participate in their seminars and helped me refine many of my ideas. Róbi Győri and his cohort of faculty and graduate students at Eötvös Loránd University made me feel at home in Budapest, taking me out for cake,

Acknowledgments

meals, and tours of the city. Bernard Hammer at the Zentrum für Umweltgeschichte and the Holbrows of Bloomington, Indiana, gave me fruitful spaces to write and revise this work. Thanks to friends in the MAGES program who have heard me talk about this project for over a decade; your kindness and laughter have made the process enjoyable. And lastly, to dear friends and family, who may not know exactly what this book is about but who always demonstrate encouraging levels of excitement and interest whenever I talk about it, thank you.

THE DANUBE EMPIRE

Introduction

THE DANUBE EMPIRE

On June 10, 1879, the venerable Hotel Imperial in Vienna hosted a formal dinner to celebrate the establishment of the Donauverein, or Danube Association. Members had formed the new advocacy group to educate citizens about the benefits of river engineering works and to petition local, national, and imperial governments in the Habsburg Empire to fund these works. In the months leading up to this meeting, local assemblies had gathered to discuss goals and possibilities for such an association and had attracted multifaceted interest groups from milling unions to agricultural lobbyists.[1] The association's hope was to transform rivers and waterways throughout the empire to facilitate transportation, promote trade, and make rivers safer and more useful to people living along them. The Donauverein's secretary, the famed geologist Eduard Suess, had overseen the Danube's regulation at Vienna a few years earlier and expressed another lofty goal for the Danube's regulation, stating that it represented not only "great cultural progress" but "a new moment in the peaceful development within the monarchy."[2]

Introduction

Figure I.1. Hotel Imperial (circa 1900) © Wien Museum.

The gathered participants at the Hotel Imperial (fig. I.1) certainly gave Suess reason to hope that the multinational Habsburg Empire, a state twice the size of modern Germany with thirty-nine million people and a dozen (officially recognized) languages, could enjoy "peaceful development" and move away from the acrimonious political rhetoric that was coming to characterize parliamentary proceedings and certain provincial politics.[3] Rather than national divisions, the Danube Association's members believed in a cause that relied on transnational unity and cooperation. The diverse coalition that had assembled in the Hotel Imperial promised to fulfill this vision. Delegates rose and spoke passionately about the Danube's unitary place in the empire, the river's indifference to tribal or national differences, its universal threat in the form of flooding, and the common call for its regulation among the populace everywhere. One participant declared that the envisioned regulation works truly fit the Emperor Franz Joseph's motto "viribus unitis" (with united strength) because they would promote the well-being of the empire's entire population.[4] Such sentiments

Introduction

imbued the Danube with a certain quality, one that had the ability to unite people through common experiences and expectations. This was certainly on the mind of the attendees, who included representatives from several riparian cities in the empire's Austrian and Hungarian halves, river engineers, assorted envoys from manufacturing and trade organizations, imperial ministerial officials, and invited guests.

When Vienna's mayor stood up to speak, he too emphasized the transnational dimension of these river engineering projects, asserting, "We would like nothing more than for all areas of the monarchy to recognize that this undertaking is not a local one, nor one simply within Austria, but rather its success will bring great fruit to both halves of the Austro-Hungarian Empire."[5] He ended his speech by toasting the Hungarian representatives present, whom he hoped would continue to spread the association's agenda in Hungary.

Government officials, prominent figures, and members of the public in Hungary appeared supportive of the Danube Association in its early days. Returning home from the opening dinner, Pál Bacsák, the vice deputy (*alispán*) of Pozsony County nestled thirty miles downstream from Vienna, presented the dinner proceedings to his constituents. He assured listeners that members of the Imperial Diet in Vienna had already promised a "warm reception" for his plans to better coordinate regulation efforts between engineers, businesses, and government agencies and that the Pozsony Chamber of Commerce planned to petition the central government in Budapest to follow suit.[6] When the Danube Association organized an excursion to travel to the miles-long rocky rapids at the empire's southeastern border, the Iron Gates, the Hungarian Ministry of Public Works and Transport likewise dispatched a representative to accompany them.[7] The trip was also opened to select members of the public, and because so many petitioned to participate the newspapers published articles declaring that the association was no longer considering attendees.[8] When some in Hungary questioned this cooperation with the "Viennese" Danube Association, others leaped to its defense, offering full-throated approbation of the association's goals of improving trade along the Danube.[9] Indeed, the esteemed Hungarian geographer János Hunfalvy lent his support to the cause, arguing that such an association and its work regulating the Danube would ultimately benefit Hungary.[10]

Regulation work was certainly not without its detractors, nor was it always smooth sailing. Nevertheless, the following chapters trace efforts by groups and individuals who pursued and undertook hydraulic engineering works as part of their vision for improving the security and prosperity of members of the public and of the empire as a whole. Improving navigation guaranteed cities and towns a reliable flow of food, merchant wares, wood, construction materials, and coal.[11] The empire's food supply strained under its eighteenth-century population boom, and reclaiming alluvial floodplains provided more arable land to feed people, a need underscored by several eighteenth-century famines.[12] Raising and unifying embankments and levees promised to protect communities from floods, while reclaimed land enabled the creation of new urban districts for housing and employing growing industrial populations.[13] Commercial infrastructure facilitated trade up and down the river and around the empire.[14] New sewage systems used the river to wash away industrial (and human) effluence, increasing sanitation when water levels were high enough.[15] Crown Prince Rudolf would say of these efforts in the late nineteenth century, "For decades, the Danube's regulation [has become] one of Austria-Hungary's most important economic [*volkswirtschaftlich*] tasks."[16]

Constructing the Danube Empire

During this period, the transformation of the Danube River's environments was often tied to new behaviors and activities that contemporaries hoped would support the well-being of the general population and underpin the functioning of the Habsburg Empire. Members of the Habsburg dynasty became deeply involved in this process, calling for the reengineering of its imperial environs to produce the desirable material conditions necessary for ensuring prosperous trade relations, stable political developments, harmonious social interactions, and strong military capacities.[17] Aside from the political utility of these engineering works, Maria Theresa and Joseph II explicitly saw the results as supporting their people's well-being.[18] Successive rulers mobilized bureaucrats and engineers and cooperated with private industries to achieve these goals. The government had no single avenue for achieving these aims, though as the opening vignette reveals, engineering the empire's waterways, especially the Danube, provided rulers and citizens with several attractive results. In the mid- to late eighteenth century, mathematicians, military engineers, and particularly those studying applied fields like "mechanics" or "hydraulics" (so-called *Hydrotechniker*) undertook engineering works on rivers that included blasting physical hindrances

Introduction

like rocky cataracts, straightening rivers by digging transections, draining adjacent marshes, and erecting embankments, all in the name of reclaiming land, preventing floods, and promoting trade.[19]

Although advocates for river engineering works frequently promised economic gains, governmental and associational representatives also emphasized desirable social and political outcomes, such as greater imperial unity. While the émigré historian Oszkár Jászi, a former civil servant in the empire, looked back a decade after the empire's collapse and bemoaned that the Danube's geographic orientation stymied integration in the empire, most nineteenth- and early twentieth-century observers painted a much brighter picture of the Danube's integrative role for state and society.[20] Early nineteenth-century travelogues frequently described the river's unifying role for the many people living along it. "Of the [empire's] rivers, we must mention above all the Danube ... with its tributaries, it encompasses two-thirds of the monarchy."[21] Another declared "that mighty artery [the Danube] arises in the heart of Europe, and through its noblest parts flow elements of life and prosperity; the wide, richly blessed valley of the mighty stream has become home to twenty-one peoples [*Völker*]."[22] Lajos Kossuth, one of the leaders of Hungary's 1848–49 uprising against Habsburg rule, perceived a natural connection among people living along the river. He envisioned the establishment of a Danubian Confederation (albeit in opposition to the Habsburg Empire) that united the different nationalities within its basin.[23] Overseeing Danube regulation works in the 1850s, the noted statistician and historian Carl Freiherr von Czoernig declared confidently, "The Austrian Empire has more advantageous water networks through its natural position than any other state on the European continent ... even if the Austrian Empire didn't have all these connections, the Danube—Europe's most beautiful and powerful river flowing through the entire length of the empire with its hundreds of miles of navigable tributaries—would still position Austria as first rate among the European states for world trade."[24] The Austrian botanist and lawyer Adolf Dürrnberger waxed lyrically that "when we observe the Danube, we feel something of its great past, and it is as if this lonely river was aware that for millennia it had been a route of world-shaping events, the carrier and communicator of Western culture, the natural founder of a great empire."[25] With such pervasive sentiments, it is unsurprising that contemporaries frequently dubbed the Habsburg state the "Danube Empire" (fig. I.2).

Introduction

Figure I.2. The Habsburg Empire with the Danube River, Major Tributaries, and Large Cities © Josh Fangmeier, 2024.

Much of the following work focuses on the ideas, plans, and efforts to engineer the river and construct this "Danube Empire." Technical experts advanced many different plans as their understanding of hydraulics evolved throughout the nineteenth century. Different interest groups also sought divergent outcomes from the engineered river, which led some to prefer certain ideas and visions over others. As the Danube was a shared space, usage and modification of the river drove negotiation and compromise between communities, individuals, and companies, geographically dispersed riparian towns, and most critically between citizens and their governments. These interactions, negotiations, and experiences along the river constitute the heart of this story.

There were, however, plenty of disagreements over the river's usage, sparking conflict, resistance, and loss. Bertalan Andrásfalvy has

Introduction

Figure I.3. View of the City Vác with Ship Mills on the Danube (1826) © Wien Museum.

detailed the ways that peasants in the Sárköz region of Hungary protested eighteenth-century efforts by landowners and royal authorities to embank the Danube. As population growth drove up the price of grain, plans to drain wetland commons and reclaim the land for private crops benefited landowners. Peasants feared—rightly, as it would turn out—that cutting the river off from its floodplain and building ad hoc embankments on parts of the river would exacerbate flood dangers and threaten land and waterscapes that peasants depended on for pastures, orchards, and fishing.[26] Tensions also arose between different professions on the river, some of whom benefited from Maria Theresa's and subsequent monarchs' championing of navigation and trade over other livelihoods. Decades of edicts regulated, reduced, and eliminated ship mills, which, for example, were blamed for worsening floods (fig. I.3). Because ship mills anchored and ground grains at the deepest and fastest-flowing parts of the river, they blocked the

Introduction

path of ship captains seeking to avoid running aground on sandbanks and shoals.[27] They were sacrificed for the sake of river commerce. The river's use was also a cause of conflict among terrestrial parties. Establishing factories along rivers to take advantage of them for transportation, cooling, and effluence disposal, manufacturers ran afoul of agriculturalists over prescribed limits of water usage and other water rights.[28]

These challenges had at their core a concern with the physical and ecological consequences of regulation and the effects thereof on humans and their livelihoods. Regulation and drainage companies that lopped off over one hundred meanders from the Tisza River and reclaimed hundreds of thousands of acres of land in Hungary got rid of the marshes, fens, gallery forests, and meadows that peasants had depended on centuries.[29] Draining wetlands reduced the diversity of flora and fauna that reed cutters, cattle farmers, net repairers, fowl hunters, and others depended on. Local fishermen and angler clubs also complained that regulation works and steamboats reduced their catches. Closing off shallow, slower-flowing side branches and deepening rivers (thus accelerating their currents) did eliminate the ecological conditions that certain fish species required to feed and breed.[30] However, speaking to the Association for Lower Austrian Geography in 1871, the founder of Vienna's Zoological and Botanical Society, Georg Ritter von Frauenfeld, stated more judiciously, "While steam navigation has caused great damage to fish stocks, none have disappeared because of it."[31] There was blame to go around; fishermen for overexploiting rivers to supply market demand, steamboats and regulation works for disrupting habitats, and factories for spewing effluents into rivers.[32] Such complaints did little to deter plans for the Danube or other rivers' regulation, which were more often than not delayed by financial or political, rather than ecological reasons.

Indeed, on the other hand, many citizens, local businesses, nongovernmental actors, and municipal authorities in the nineteenth century were actively involved and invested in this process of physically transforming rivers around the empire. Much the way the Danube Association envisioned, these groups used governmental and democratic avenues to engage with provincial, national, and imperial authorities. They used these channels to assert their interests and to shape the visions set out by central governments so that hydraulic engineering works would protect their material well-being, not threaten it. Trans-

Introduction

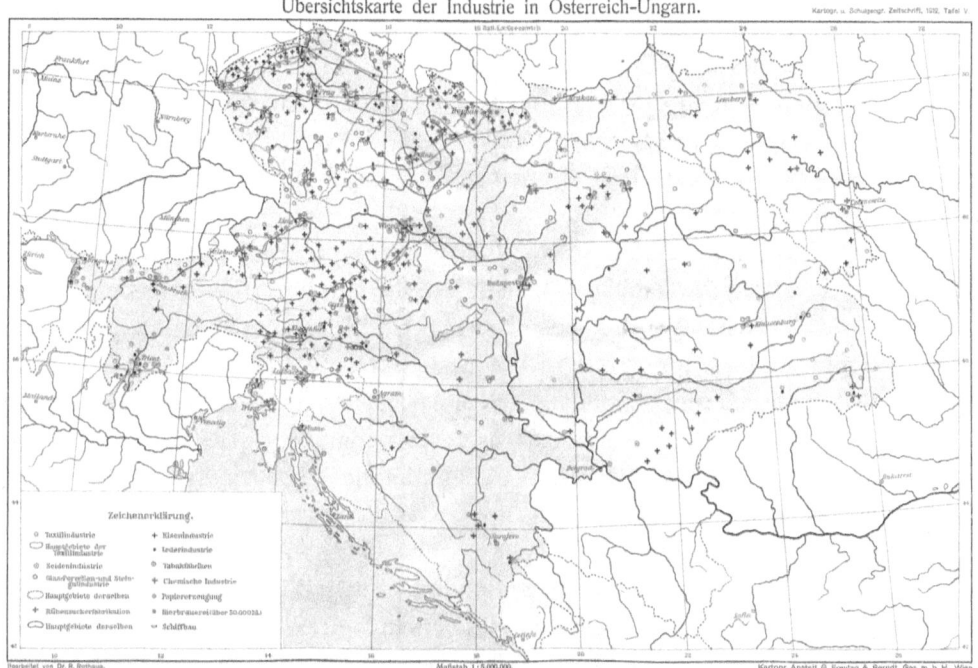

Figure I.4. *Overview of Industry in Austria-Hungary,* 1912 © freytag & berndt-brands.

forming the Danube changed people's practices and engagement with each other throughout the empire, reflecting the Danube's transition from a local, communal space into a more cohesive, imperial realm.

The Danube Question

The Danube River's prominence provided a clear focal point to channel interest and support from inhabitants of the empire for its regulation. The river ran for more than 850 miles through the heart of the empire, and its major tributaries knit together Alpine territories in the west, the lower Bohemian Massif in the north, and nearly the entire Carpathian Basin in the south and east. When the Danube Association constituted itself, the empire encompassed 5,000 miles of navigable waterways. Steamboats only plied one-third of Austria's navigable waterways and three-fifths of Hungary's, while the remaining waterway traffic was confined to rafts, barges, and other smaller vessels.[33] These waterways were useful for moving around industrial, commercial, and agricultural goods, but they also did all sorts of work, from driving machines to cooling industrial processes. Imperial geographers clearly illustrated the dependence that such sectors had on rivers. As part of an educational manual for young students, Rudolf Rothaug's cartographic "Overview of Industry in Austria-Hungary" revealed at a glance the dense cluster of industries situated on the empire's major river systems (fig. I.4).

Although the Danube's expansive geography loomed large in domestic affairs, the Danube Association's adherents nevertheless recognized the challenges that existed to making the river system a more significant force for empire-wide cohesion. Despite statistical reports that published precise total lengths of navigable rivers each year, navigability on the empire's rivers was never a constant figure. Instead, it was subject to annual and regional climatic shifts. Unseasonable flooding and drought affected shipping, as did seasonal variations in high and low water levels brought on annually by winter freezes, glacial melts, spring and summer rainfall, and dwindling precipitation in late fall and winter.[34] The complex climatological influences on the Danube catchment area—rainier marine in the west, drier continental in the east, and milder Mediterranean in the south—meant that precipitation and water levels were hard to predict from year to year.[35]

With such variable water levels, flood prevention was also a key agenda of river regulation works. Flooding had both natural and anthropocentric causes, as chapter 4 will explore in greater detail. Riparian communities around the empire were subjected to an inordinate number of summer floods and winter ice-dam flooding on the Danube and on other rivers around Europe in the eighteenth and nineteenth centuries thanks to the final iterations of the Little Ice Age, which brought colder winters and rainier summers to parts of Central Europe.[36] Human activity also exacerbated conditions that led to flooding. Early modern deforestation in the Danube Basin, changing land usage patterns along the river and its tributaries, and the empire's growing population also disrupted the soil and accelerated the erosion of riverbanks.[37] These factors caused rivers to meander, become shallower, and more readily freeze and flood. Engineers and companies dredged and deepened rivers, channelized branching arms into a single bed, and constrained rivers behind embankments and levees in a grand effort to minimize the chance of flooding, but these attempts were not always successful.[38] Mindful of the intricate and interlocking factors governing the Danube's hydrology, the Hungarian journalist Kornél Szokolay succinctly summarized this challenge for his readers in 1880, arguing, "We can only start regulation when a plan for the entire Danube is established, because we cannot improve certain stretches without unregulated portions deteriorating further."[39] Overcoming the diverse conditions along the empire's waterways demanded unity of vision.

These challenges were part of a broader "Austrian Question" that Deborah Coen has described in *Climate in Motion: Science, Empire, and the Problem of Scale* regarding strategies of governance that went hand in hand with management of the environment. Coen argues that governing cultural diversity in the empire mirrored the immense logistical challenge of governing an empire filled with vastly divergent climates and topographies, from semiarid steppes to cool mountain pastures to flood-prone alluvial plains. Contemporary scholars, naturalists, and officials around the Habsburg Empire recognized this difficulty and endeavored to study and understand the interrelated components of these diverse natural features. Armed with this knowledge, they devised strategies to both address the challenges they presented and expound the many virtues and advantages that such varied landscapes and climates conveyed to the empire and its people.[40]

The Danube Association also tried to mobilize action to address these common challenges. Promoting the river's regulation to both advance navigation and reduce flood risks was part of its empire-wide activities. Its influence uniting political and budgetary agendas in legislatures around the empire was palpable in the decades following its establishment. The association also gained imperial patronage from Crown Prince Rudolf and later boasted Franz Ferdinand's support as well. Beyond organizing speaking events and petitioning the government, the association's leadership also organized special excursions up and down the Danube to bring government officials and members of the public on steamboats to inspect firsthand ongoing regulation works and to visit unregulated stretches that local communities wanted help modifying. The influential *Wiener Presse* credited the association with inspiring a resurgent interest in the Danube Question.[41] Advocates petitioned Austria's Imperial Diet (Reichsrat) on behalf of provincial interests, and the Danube Association's publications influenced engineers and representatives in both Austria and Hungary.[42] Public engagement coupled with government directives ensured funding for the Danube's regulation for the remaining decades of the empire's existence until 1918.

Coal, Steam, and Industrialization Fuel Dominion over the Natural World

Riparian communities in the Habsburg Empire were not alone in wanting to redefine their relationship with nature. Until the modern era, natural elements like disease, climate, and soil nutrients strongly influenced all human development, limiting population growth, food production, and economic output. Fernand Braudel consciously compared these constraints to Europe's onerous socioeconomic and political burdens when he labeled them the "Biological *Ancien Régime*."[43] Like Hercules battling against the shape-shifting river god Archelous to win Deianira's hand in marriage, however, humans recognized that to battle and subdue nature and to overcome its constraints would grant them the desired goal of a more amendable (and profitable) environment within which to live.

The energy to accomplish these tasks was, for most of human existence, limited to work done by human and animal muscle, which was fed a steady diet of carbohydrates that plants had photosynthesized from the sun's energy. Climate and geography affected the solar energy and precipitation available to plants and thus the calories that humans could harness to do their work. To supplement their labor, inventive humans built mills to capture wind and water—fluids that move and circulate due to the sun's energy—or devised technologies for making human and animal labor more efficient.[44] The sun also powered the growth of forests and grasses, which humans burned for heat to cook, keep warm, and make all sorts of things from soap to ceramic. Coal, the fossilized remains of organic matter, also served this purpose where it was available.

The herculean task of remaking nature gained powerful tools with the harnessing of steam power during the Industrial Revolution. The knowledge necessary to build and operate steam engines was honed over a century of observations, devices, and discoveries. By the early eighteenth century, Thomas Newcomen's atmospheric engine burned coal to boil water and release steam into a cylinder, which, when cooled, formed a vacuum, allowing atmospheric pressure to depress a piston into the cylinder and draw water up a pipe. This invention freed coal from its use as an ersatz fuel in place of biomass and instead used it

Introduction

to accomplish tasks that had hitherto been the primary realm of muscle power (or, in limited instances, wind and waterpower). Early steam engines were admittedly quite inefficient, and their main profitable application was pumping water from coal mines to free up waterlogged coal seams. As steam engines proliferated at pitheads around Europe, the cost of coal decreased. With cheaper coal came the classical litany of inventions and discoveries that defined industrialization: heavy industry, and speedier manufacturing, communication, and transportation.[45] By harnessing coal, humans intensified their ability to extract, exploit, and remake nature according to their own designs.

The transition to burning fossil fuels restructured the nature of social, economic, political power around the world. Andreas Malm argues that the etymological overlap in English with "power" denoting both a physical, natural phenomenon *and* the human-oriented notion of dominion or authority does not occur in other languages but that this conflation is apt as "the power derived from fossil fuels was dual in meaning and nature from the very start."[46] Emerging industrial nations shifted geopolitical power away from earlier global industrial and commercial powers, China and India, and fueled Europe's influence over other parts of the earth.[47] By the late nineteenth century, a Second Industrial Revolution, characterized not only by steam power but by heavy industry and electricity, churned out new inventions and methods for mass-producing consumer goods, synthesizing new chemicals and materials, extracting and processing natural resources, and modifying the landscape. With these inventions, Western European countries expanded and cemented their grip on overseas colonies, which they had conquered to feed Europeans' voracious appetite for raw materials and commercial markets. Corey Ross has argued in *Power and Ecology in the Age of Empires: Europe and the Transformation of the Tropical World* that European empires' political control over their colonies was integrally linked with the modification and control of local ecosystems.[48] In the Habsburg Empire, there were also clear gradients of industrialization that shaped relations between different regions. The Bohemian and Austrian lands were the first to see mechanization, connected to the textile industry, while the eastern provinces' main exports well into the nineteenth century remained agricultural products. This provided fodder for nationalists and later scholars to decry the colonial relationship of peripheral raw producers in supplying industrialized core lands.[49] Historians today acknowledge these differences in economic development but are more circumspect about overt comparisons between Western Europe's overseas colonies and "internal colonies" in the Habsburg lands. Likewise, for Malm, the greatest legacy of industrialization was not the economic dependencies or colonial relationships that it produced but the inordinate amount of carbon released from combustion that has accumulated for centuries in the atmosphere and now contributes to our drastic climate crisis.[50]

Situating the Conversation

The Danube is a fruitful way to look anew at the Habsburg Empire. Scholars of the empire have long concerned themselves with questions of identity, loyalties, and national development.[51] Many newer works, like Pieter M. Judson's *The Habsburg Empire: A New History*, argue that the state (and its rulers) were adaptable, responsive, and resourceful in the face of economic, social, martial, or nationalistic challenges.[52] Asking these questions through an environmental lens yields refreshing new insights into the empire's history. The Danube transcended national boundaries, its usage involved diverse stakeholders, and it was a dynamic product of both human interventions and natural processes. These characteristics flow counter to many traditional narratives and provide a new approach to a well-worn history.

Indeed, studying rivers has proved fertile ground for scholars to produce dozens of original histories in the past few decades. In some instances, historians use rivers or the natural world more broadly to rebalance questions of agency. These works studiously document how human actions, decisions, and behaviors are embedded in socioenvironmental spaces, some governed by humans, some by natural elements, and many by some combination of both. In his iconic work *The Organic Machine: The Remaking of the Columbia River*, Richard White uses the concept of energy to reconceptualize human relations along the Columbia River. White's work focuses on the kinetic energy of the river and thus the energy expended to ply the river (human muscle and later biomass/fossil-fueled ships), the energy of fish calories (and competing claims of those wishing to catch them), and the river's energy that fueled hydropower, irrigated the land, and powered western development, all of which involved and affected people from Indigenous tribes to immigrant laborers to rural farmers.[53] Although this book does not touch on fish or dams à la White, energy does play an important role in this story. As we will see in the following chapters,

Introduction

traditional professions on the river, such as towing crews, ship mills, raft captains, industrialists and others harnessed and combated the river's flow, while new energy sources, like steam-powered boats and excavators, became the tools and rationale for altering the river and people's relationships with it. The river's flow (its energy) grew and shrank annually depending on seasonal precipitation and snowmelt with implications for navigation, flood protection, and sanitation, all of which had an impact on communities close to the empire's rivers and occupied the minds of decision-makers everywhere.

This book, like many others, also illuminates the iterative role that rivers play in state-building processes.[54] Two works exemplify how these processes predated and postdated the Habsburg Empire in neighboring states while exploring many of the same questions of nature and governance. Faisal H. Husain's *Rivers of the Sultan: The Tigris and Euphrates in the Ottoman Empire* reconstructs the ecological and physical power of the two titular rivers and their alluvial plains, control over which underpinned the early modern Ottoman state's economic, martial, and political power. Although less eponymous than the "Danube Empire," the Ottoman Empire's rivers greatly influenced imperial governance, both strengthening and, at times, imperiling state-building efforts. Husain enriches this narrative by combining Ottoman imperial source material with ethnographic interpretations and breakthrough scientific discoveries. These provide insight into the behavior of local groups from boat makers to herdsmen, whose activities and loyalties were embedded in the local landscapes and influenced the operation and maintenance of Ottoman imperial reach along its southeastern border.[55] Sara B. Pritchard's work *Confluence: The Nature of Technology and the Remaking of the Rhône*, meanwhile, provides a detailed history of how vast technological, agricultural, and industrial interests reshaped the Rhône after World War II. Pritchard delves into the calculations of French politicians, industrialists, engineers, and local members of riparian communities whose plans for transforming the river promoted both material and cultural agendas in reconstructing the nation after the war. In her story, Pritchard depicts French state-building as a process of negotiation rather than something natural or predetermined.[56] In each instance, the Rhône and the Tigris-Euphrates Rivers are material bodies that drive these histories as much as the people within the stories do. These histories also demonstrate the extent to which central governments relied on

Introduction

local communities' cooperation to shape and advance national and imperial agendas.

The Danube is, admittedly, not a new topic. It has also been the subject of countless historians, authors, journalists, poets, and linguists for centuries. In most historical works, however, the river disappears quickly from the work's narrative or features merely as a backdrop to situate political, military, or social developments. In recent years, scholars affiliated with the Center for Environmental History (ZUG) in Vienna, such as Gertrud Haidvogel, Severin Hohensinner, Martin Schmid, Verena Winiwarter, and many others have provided a refreshing exception, placing the river at the center of their scholarship and narratives. They have been particularly prodigious in undertaking historical-ecological research, and their rich, quantitative case studies have produced numerous local and national histories of the Danube River that document the river's contribution to energy flows along it, the ecological and social consequences of its usage before and after industrialization, and its role in urban development and metabolism.[57] To augment this focus on purely Austrian stretches of the Danube, this book uses historical sources in German and Hungarian to integrate the long stretch of the river from Engelhartszell to Orsova. Gone is the conventional periodization of the empire that studies "Imperial Austria" independently of "Royal Hungary" after 1867. This book instead tells a shared story about both halves of the empire united by the Danube River.

Structure

Chapter 1 explores different facets of the Habsburg dynasty's efforts to create the "Imperial Danube" as a means to legitimate its rule and foster the well-being of the empire's population. It studies the dynasty's efforts to create a modern state apparatus to help it modify and "control" nature, the ways the dynasty and its bureaucracy supported technology and innovation to support the populace's economic interests, the dynasty's efforts to appear the traditional "protector" of its people by safeguarding their material well-being through natural disasters, and the modern methods the dynasty employed to retain citizens' loyalty through patronizing and celebrating public works along the river.

Chapter 2 looks at the evolving negotiations between all levels of government and the public to make the Danube and its tributaries a reliable avenue for commerce and transportation or a "life artery," in

Introduction

contemporaneous parlance. The chapter argues that few permanent river regulation works were possible in the preindustrial era, when provincial governments and local communities retained their traditional responsibility for undertaking hydraulic engineering works. This is followed by a description of the turning point when mid-century flooding and empire-wide uprisings led to political changes that streamlined and centralized the funding and planning of engineering works. The chapter then argues that members of the public successfully promoted local environmental interests by working within the extant instruments of civic engagement (petitions, elections, and lobbying), thereby reifying structures of empire. Finally, the chapter examines how ongoing interests in expanding canals and waterways often required local, national, and transnational cooperation between the empire's Austrian and Hungarian halves, despite antagonistic nationalistic rhetoric that was common among politicians.

Chapter 3 argues that the hydraulic engineering projects undertaken on the Danube, described in chapter 2, promoted greater social and economic integration within the empire. The chapter provides an overview of the traditional traffic on the river before studying how steam navigation became a forceful reason and tool for engineering the river. The chapter analyzes hydraulic engineering expenditures, passenger data, and freight records to quantitatively demonstrate that both steamboats and traditional traffic on the Danube and its tributaries kept waterways a commercially integrative force between the two halves of the Habsburg Empire. Finally, looking at data from riparian communities, the chapter argues that urban growth drove Danube traffic and helped integrate the empire's hinterlands into the Danube economy. These conclusions complicate traditional assumptions about "national loyalties" in the empire by demonstrating how the river forged transnational dependencies and reified imperial economic and commercial connections even up to 1914.

Chapter 4 argues that natural disasters and environmental concerns helped forge modern aspects of imperial rule. It maintains that the shared experience of flood disasters created transnational networks of solidarity among the populace, as donors and helpers from around the empire routinely aided flood-stricken communities, despite vast geographical distance. Empire-building also expanded in the wake of floods and natural disasters, prompting the foundation of new governmental and societal mechanisms for flood relief and disaster prevention. These measures relied heavily on centralized planning and public investments as well as on cooperation between imperial and local initiatives to advance the "common good." Finally, the chapter argues that empire-wide flood prevention measures were only the physical side of protection, and that imperial and local governments worked with volunteers, associations, and committees to harmonize behavior and practices on the river to minimize flood dangers and optimize assistance during floods to ensure the well-being of all.

Chapter 5 studies how local communities negotiated the evolving usage of communal spaces along the Danube. It examines how reclaiming flood plains and embanking the Danube created new spaces along the river for use in industry and trade, public health, and recreation. Local regulation works and infrastructure projects tied people's well-being to the river and, due to central planning and funding, strengthened their everyday connections with imperial authorities. A case study of two cities, Linz in Austria and Győr in Hungary, reveals the ways that municipal authorities helped navigate common usage of the Danube River among many stakeholders. Those stakeholders, for their part, channeled concerns, petitions, and speeches—individually or within an association or business—through extant power structures, such as local city councils, provincial diets and governors' offices, and ultimately also imperial diets and ministries. This civic engagement indicated their expectation (and trust) that existing authorities were responsible for and responsive to their requests. The French historian Ernst Renan asked, "What Is a Nation?" and answered that it was a daily plebiscite of people's everyday actions that collectively reaffirmed their commitment to the nation.[58] If this is the case, then civil and governmental actions collectively affirmed their support for the state, and these interactions, activities, and behaviors provided crystal-clear evidence to answer the question: "What is the Danube Empire?"

This book presents the nineteenth- and early twentieth-century social and material conditions that influenced people's behavior and interactions along the Danube and seeks to understand how the river served as a source of loyalty and cohesion in the empire. For the authorities, the river's natural connections throughout the empire and the ubiquity of its presence in many people's daily lives made the Danube a natural place to mediate interactions and to engineer the river in a way that protected citizens, improved their material well-being,

Introduction

and channelized loyalty to the dynasty and imperial state. Citizens were actively involved in this "state-building" process. They turned the imperial government's attention to local needs, relying on the appropriate structures of interaction between "state" and "society"—political petitions, associational movements, elections, and representation—and, in rare cases when those did not suffice, on agitation and revolution. They were flexible and pragmatic in pursuing local needs, knitting together governmental and commercial support for their endeavors. The Danube was instrumental in this process. Its nature and behavior represented both threats and opportunities for the empire's inhabitants. Through its wide-ranging influence on daily life, the river in gradual and sudden ways compelled governments, businesses, and people to consider it as an imperial issue, something that required the empire's entire population to address and overcome it, in the motto of Emperor Franz Joseph, "with united strength."

Chapter 1

CREATING THE IMPERIAL DANUBE

In October 1823, a newly completed steamboat, *Franz I*, departed its harbor at Fischamend and paddled a few miles up the Danube to the Austrian imperial capital Vienna for public viewing at the Prater gardens.[1] The boat's presence promised to be a curious event for residents, as for most subjects in the Habsburg Empire, as the empire's few steam engines in operation were confined to Moravia's textile industry.[2] With paddles rotating on either side and its movement upstream against the current, *Franz I* contrasted starkly with the Danube's typical river traffic arriving by the thousands each year in Vienna: flat-bottomed barges bearing everything from salt to stone, large rafts destined to be broken up for firewood, and hired passenger boats of various sizes. The *Franz I* was commissioned by two Frenchmen, who had already built and sailed a steamboat on the Danube several years earlier between Pest-Buda and Komárom (the Duna in 1817) and were hoping to capitalize on this moderate success with a new steam navigation company.

Figure 1.1. Josef Hoffmann, *The Danube Floodplain at Vienna*, 1870 © Wien Museum.

The new steamboat faced many hindrances. The Danube at Vienna flowed with the speed of a mountain stream and fanned across the Viennese floodplain in arms of varying and variable depths (fig. 1.1). Low water levels in the arm adjacent to Vienna had already forced the entrepreneurs to abandon plans to show off the boat near the fashionable horse tracks located, according to one observer, where the city "rests on the waters of the Danube, the foliage of the Prater, [where] the merry crowds [stream] along to enjoy its shades."[3] Relocated to another landing place farther downstream, the berthed boat was nevertheless festively celebrated by members of the public, the empress, numerous archdukes, and even the son of Napoleon Bonaparte, Emperor Franz's grandson, who came by to inspect it. After its reception in Vienna, *Franz I* took off downstream to Pest for additional festivities hosted by Franz's brother Archduke Joseph, the popular palatine of Hungary.[4] Additional difficulties arose on the boat's return trip to

Vienna. A constant battle against the river's current, stopovers to replenish the boat's fuel supplies, misty conditions, sandbanks, storms, pontoon bridges, and low water levels all delayed the boat's progress. Only on the tenth day did it finally arrive back in Vienna. Disregarding these setbacks, newspapers reported the trip a success, remarking how impressive it was that the boat managed the journey in late autumn when meteorological and hydrological conditions were particularly unfavorable.[5]

What was the significance of this trip? Frank Uekoetter argues that environmental historians identify turning points in history, because they "provide a backbone to narratives that no scholarly study of history can do without."[6] Was this voyage such a moment? It was not. *Franz I* went on to be a commercial failure, its makers having been unable to establish regular schedules or make the enterprise profitable. Instead, *Franz I*'s journey indicated a transitional moment in the relationship between humans and the environment, one in which humans were beginning to mobilize newer tools, like steam engines, to gain the upper hand over the forces of nature though often still with limited success.

If steam engines were only slowly reorganizing human relationships, it was significant that they arrived just as political developments were violently challenging Europe's status quo.[7] By letter and by army, the Enlightenment, the French Revolution, and twenty-five years of European-wide warfare were threatening to sweep away the traditional powers and privileges of the ancien régime. Even after these subsided, revolutionary ideas continued for decades to churn uprisings against feudal restrictions and aristocratic and monarchical privilege. Coming to the throne in 1792, Emperor Franz II immediately experienced the brutal manifestations of nationalism and liberalism: the bloodshed of the French Revolution, which claimed the head of his aunt, Marie Antoinette, and a quarter century of warfare against French armies and their allies. Regarding these ideologies with natural suspicion, Franz nevertheless passed many laws that adhered to the spirit of his predecessors' enlightened reforms, governing to the benefit of his subjects.[8] In this new political environment, the emperor's actions and behavior took on great significance as they underpinned his authority and legitimacy, much as notions of divine sovereignty had for other monarchs in the not-too-distant past.

As this chapter explores, these transformations both environmental and political converged on the Danube River, where the dynasty mobilized its power to control nature and legitimate itself. Successive Habsburg monarchs took steps to engineer the river in a way that projected imperial power, inspired devotion to the dynasty, and, where possible, united its diverse population. To do so, the monarchy first assembled the bureaucracy and engineering corps needed for carrying out hydraulic engineering projects on the Danube. Its policies then actively supported steam navigation's expansion and more cohesive trade within the empire. Finally, successive Habsburg monarchs, like many of their contemporary rulers, staged celebrations along the river to demonstrate their continued attentiveness to securing the well-being of their people.[9] By investing in and celebrating along the Danube, the Habsburgs made the institution of the monarchy more visible and more accessible to their people, providing ample opportunity for spontaneous—and staged—expressions of devotion to dynasty.

The Danube and the Dynasty

Eighty years before the *Franz I* set sail, twenty-three-year-old Maria Theresa ascended to the Habsburg throne after her father, Charles VI, died unexpectedly. Having been kept from the responsibilities of governing, Maria Theresa was unprepared in one sense to govern the large, diverse territory that she now found herself at the helm of. The state's immense size was, in part, thanks to military victories of her father and grandfather in the seventeenth and eighteenth centuries, which had liberated Hungary, parts of southeast Europe, and Transylvania from the Ottomans after 150 years' occupation. Less fortuitous than her predecessors, Maria Theresa ascended the throne and promptly lost her most valuable province, Silesia, to her rapacious neighbor, King Frederick II of Prussia. The War of Austrian Succession (1740–48) and the Seven Years' War (1756–63) failed to wrest back control of the province. Facing immense challenges both at home and abroad, Maria Theresa spent most of her rule introducing economic and educational reforms and consolidating the empire's political institutions to strengthen the state's financial health and military capabilities.[10]

The Danube figured prominently in these reforms. Her father and grandfather's military victories had more than doubled the length of Danube under Habsburg control and brought the river-rich Carpathian Basin back under Habsburg rule. Unfortunately, centuries under Ottoman occupation had wreaked havoc on the region's hydrological conditions. Commerce on the Danube had stagnated by the seventeenth century due to depopulation, changing consumption habits upstream, and the devastation of taxes and dues exacted from the peasant population to wage war. Peasants retreating from the invading Ottoman army had left their fields, either settling in towns for greater protection or fleeing to the small strip of territory, "Royal Hungary," that the Habsburgs had retained. In the absence of regular farming, land along the Tisza River reverted to marshes and became malaria infested, plaguing inhabitants until the nineteenth century.[11] To address these conditions, Maria Theresa implemented policies to improve navigation and trade, protect subjects from flooding, and repopulate the emptier parts of her realm with colonists, many of whom sailed down the Danube to settle along the fluvial borderlands with the Ottomans.

David Blackbourn has pointed out that projects like river regulation and wetland drainage works "bore the hallmark of cameralist science, which called for maximizing resources," and were not unique to the Habsburgs.[12] In his seminal work *The Conquest of Nature: Water, Landscape, and the Making of Modern Germany*, Blackbourn explores how Prussia's Frederick II employed such projects in the mid-eighteenth century as part of a dual process of conquering land, which both tamed nature and made it (and its human inhabitants) more productive. Such projects required the absolute monarch to overcome locals' objections, often by military force, while he mobilized resources, experts, and workers to subdue nature in the same way that his armies pacified his enemies.[13] For rulers in Central Europe who lacked resource-rich, overseas colonies, cameralist policies provided an attractive and—more important—a feasible approach to strengthening the state.

Another influential philosophy that promised to generate immense national wealth was open or free trade. In a talk given several decades before his *The Wealth of Nations* (1776), Adam Smith stated succinctly about state wealth: "Little else is required to carry a state to the highest degree of affluence from the lowest barbarism but peace, easy taxes, and a tolerable administration of justice." Smith likened taxes, customs, tariffs, and any other obstacles to trade to a government diverting the "natural course" of human affairs into a restrictive "channel."[14] This ethos was at the center of certain pronouncements from the American Founding Fathers, who wished to increase national economic unity. Clearing their domestic waterways from any hindrances to trade, representatives declared in the 1787 Northwest Ordinance that "the navigable waters leading into the Mississippi and St. Lawrence, and the carrying places between the same, shall be common highways and forever free . . . without any tax, impost, or duty therefor."[15] Free trade also animated rulers in the Habsburg state. Maria Theresa's father, Charles VI, had pursued many policies to enliven trade: reducing and eliminating certain dues between provinces, building a road from Vienna to the Adriatic port city Trieste, declaring Trieste and Fiume "free harbors," and establishing companies designed to take up trade in the Black Sea, Mediterranean, and the East Indies. Following his lead, Maria Theresa expanded road networks and built canals, supported trade companies, established a Mediterranean flotilla, and abolished "illegal river dues" from the Danube.

But Habsburg leaders followed the free trade maxim only insofar as it benefited national/imperial interests. Maria Theresa was no exception. When she convened the Hungarian Diet in 1751 to discuss Hungary's military and tax contributions, the Hungarian aristocracy refused to relinquish their tax-free privilege, pointing to the ongoing economic underdevelopment resulting from the Ottoman occupation. The monarch responded by establishing an internal customs border and raising tariffs on Hungarian (mostly agricultural) products to raise funds in lieu of taxes. Tariffs and customs lopped off Hungarians' access to German markets and their traditional sources of credit. Maria Theresa combined this with policies that removed economic incentives for expanding manufacturing and trade in Hungary—available to the Hereditary and Bohemian lands—which hampered the foundations for Hungary's economic recovery.[16] Despite these barriers, Hungary and Austria enjoyed a preferential trade arrangement up until the customs union in 1850. According to John Komlos, "Each partner paid a lower duty on goods originating in the other's territory than on those from an outside source."[17]

Such policies were not alone in influencing the empire's internal trade; so too did the physical conditions of the empire and its infrastructure. When a famine struck in 1770, for example, navigation on the Danube was so difficult that repeated shipwrecks threatened grain shipments to hungry regions.[18] The next year Maria Theresa established the Navigation Directorate with the avowed purpose of improving shipping conditions on the Danube and its important tributaries.[19] Maria Theresa entrusted her faithful adviser Franz Anton von

Raab, architect of the successful Robot reform in the Bohemian lands, to oversee plans to improve navigability and reduce flood dangers in Hungary. The directorate's agents found that ship mills were some of the greatest culprits to navigation. They physically blocked boat traffic and changed the river's hydrology, eroding banks, forming new sandbanks, and forcing the river to meander and grow shallower. They were also ubiquitous; over five hundred mills were registered on the Danube between Pozsony/Pressburg (modern Bratislava, Slovakia) and Semlin (modern Zemun, Serbia). Writing back to Vienna, agents complained that millers disregarded their directives to move their boat mills to side channels.[20] This was no surprise to the administration. In 1764, when Maria Theresa's cabinet secretary, Cornelius Neny, had solicited opinions about what most concerned the general population, a key member of the Hungarian Lieutenancy Council, Ferenc Balassa, replied, "It would be extremely beneficial to the common good [*dem gemeinen Besten*] . . . to eliminate mills found in the Danube, Drava, Sava and other navigable rivers . . . [that] are placed into the water with rods, which remain in the water over the winter and rot, [and] cannot be drawn out again. The river floods its banks, shipping is impeded, and fishing is completely spoiled."[21] In keeping with this conclusion, Balassa advanced proposals for the river's usage that prized navigation and reduced flood dangers. Rivers were to be cleared of hindrances and mills were to be moved to smaller streams that were neither navigable nor in any danger of flooding. In this way, navigable rivers could be left free for ships. It is ironic that Balassa blamed millers for threatening fishing practices, given that the following century most fishermen would blame river engineering works and the rise in river traffic for their dwindling catches.[22]

Mills were not just inconvenient but potentially dangerous. The Hungarian novelist Mór Jokai penned his 1872 work *Az Arany Ember* (The man with the golden touch) about grain traders on the Danube in the 1830s. In the opening scene, the boat of captain Mihály Timár is being towed past the dangerous Lower Danube rapids when an unmoored water mill comes careening toward him and his passengers "sweeping away the mills it met on its way, and sinking any cargo-boats which could not get out of its path." Trying to avoid losing control of his ship in the swirling rapids, the boat captain uses a boat hook to maneuver the oncoming mill into a turbulent eddy nearby, sinking it.[23] The harrowing encounter, although fictional, reflects contemporary accounts in newspaper articles, governmental petitions, and business minutes testifying to the liability of ship mills.

Aiming to clear other dangers from the riverbed, the Navigation Directorate also blasted rocks and reefs and dredged sandbanks to clear the path for boats. The most prominent obstacles that hindered navigation on the Upper Danube were the *Strudel* and the *Wirbel*, whirlpools caused by rocks along the riverbed at each location. Brewster's encyclopedia in 1830 regaled readers with descriptions of both sites:

> A dreadful noise, like that of thunder, soon announces the famous waterfall and whirlpool of Stroudel. It has frequently proved fatal to boats drawn into its vortex; but if the boatmen are not intoxicated and the water is not too low, there is no risk of any accident ... about a quarter of a league farther is the whirlpool of Wirbel, still more dangerous. The impetuous waves of the Danube, which here dash against an inclined promontory of the rock, are driven back in rapid circles across the narrow strait confined between two lofty banks. before they enter, and after they quit these two whirlpools, the boatmen regularly say their prayers.[24]

Heading up Maria Theresa's project of clearing these rapids was the Jesuit naturalist and engineer Joseph Walcher. Born in Linz in 1719, Walcher discovered a passion for mathematics and its applied disciplines. After graduating from high school, he joined the Jesuit order, where he eschewed the priesthood and instead set off on a journey through the Hereditary Lands of the Habsburg state to observe road construction projects and hydraulic machines. Settling in Vienna, Walcher later taught Hebrew and mathematics at the Collegium Theresianum while developing courses in mechanics and consulting on river engineering projects in Vienna and Hungary. With the dissolution of the Jesuit order in 1773, state authorities recruited Walcher to be navigation director in 1774, and he led efforts to clear the *Wirbel* and *Strudel* (fig. 1.2).[25] After several years' work, Walcher summarized the contradictory dangers that these two stretches offered; namely, "at the *Strudel* the largest danger is during low waters when dangerous reefs are just below the surface and the river's current can toss ships against them ... in rising waters, the danger subsided ... at the *Wirbel* [however], low water levels just produce weak eddies which even small ships can pass through without much notice. At high waters, these eddies, which make up the *Wirbel*, are so terrible and violent, that boatsmen

Figure 1.2. Jakob Alt, *The Wirbel Rapids*, 1826 © Wien Museum.

have to apply all attention and strength to overcome them."[26] Blasting a channel through the rocks at the *Strudel* rapids began in 1777 and work on both stretches lasted nearly two decades.

These early projects in the Habsburg Empire were undertaken in the same spirit as large-scale river engineering and canal building projects that rulers across the globe spearheaded to knit together their geographically dispersed peoples and economies. Writing in the sixteenth century, the Jesuit traveler Matteo Ricci wrote about China, "This country is so thoroughly covered by an intersecting network of rivers and canals that it is possible to travel almost anywhere by water."[27] In neighboring Russia, Voltaire expressed his approval for Peter the Great's ambitious works in the late seventeenth and early eighteenth centuries, claiming that the "construction of canals, which joined rivers, seas, and people, that nature had separated from each other" was one of the Russian emperor's greatest advancements for his people.[28] Frenzied commercial speculation and several parliamentary acts fueled a boom in canal stock companies in Great Britain and led a member of the Brussels Academy of Sciences and Fine Arts to rhapsodize that "the great number of extensive and magnificent canals, which have been cut through almost every part of England of late years, for the use of internal navigation... merit to be considered in a scientifical as well as in a commercial light."[29] Looking back at the early American Republic, Theodore Roosevelt claimed that the US Constitution itself resulted from the Founding Fathers' quest to resolve practical questions of federal and state power related to, among other things, interstate navigation, river regulation, and canal construction.[30] By the early nineteenth century, "German hydraulic engineers showed a confidence born of growing experience and technical expertise" thanks to regulation projects on the Oder, Warthe, Elbe, and Rhine Rivers, using increasingly refined techniques to overcome the deficiencies and inadequacies of earlier reclamation and regulation works.[31] The process of river regulation and canal construction served in each instance to advance specific interests in each state; transportation, land reclamation, flood protections, or irrigation.

To support works within the Habsburg state, Maria Theresa and Joseph II expanded the bureaucratic and technical corps responsible for executing these projects.[32] After establishing the Navigation Directorate in 1771, Maria Theresa granted it the power to oversee local regulation plans. The Hungarian Treasury and the Lieutenancy Council followed suit and set up the Navigation and Engineering Department (Hajózási és épitészti osztály) in 1785. In 1788, Joseph issued an edict establishing the General Engineering Directorate, whose overarching authority enabled it to direct engineering projects throughout the entire empire.

Undertaking the actual work of these technical bodies fell mostly on the military's engineer corps. Traditional training for these men included surveying and mapmaking, work that helped guide military decisions about provisioning troops and securing territory. According to Madalina Valeria Veres, engineers' ability to visually depict the empire in aggregate contributed to the "restructuring of the Habsburg Monarchy from a loosely connected empire to a centralized state."[33] Adding to their ranks, Joseph II established the Institutum Geometrico-Hydrotechnicum in 1782 to educate a new generation of engineers in Hungary.

Creating the Imperial Danube

Some of the Danube's most influential engineers in the following decades emerged from these propitious circumstances. The military engineer József Kiss was responsible for regulating the Danube near Pozsony/Pressburg and organizing the construction of the Franz Canal (1793–1802), which connected the Danube and Tisza Rivers. József Beszédes headed numerous regulation surveys along the Danube after graduating from the Institutum Geometricum in 1813. The Hydraulics Directorate, led by the engineers Pál Vásárhelyi and Mátyás Huszár, began regulation work in the 1820s and 1830s—the empire's prominent rivers had been surveyed after 1817. Some of the first transections, cuts in the Danube's meanders, were undertaken in this period and engineers attempted where possible to harness and guide the immense energy of the river itself to dig out and move along gravel and other material during this work.[34] Vásárhelyi was also instrumental in drafting a plan in the 1830s to remove arguably the largest hindrance to navigation along the Danube, the Iron Gates, a fifty-mile stretch of rocky rapids between the towns of Moldova and Turnu-Severin on the empire's southeastern border.[35] Much like members of the US Army Corps of Engineers cutting their teeth on levee projects along the Mississippi in the nineteenth century, by the end of the eighteenth century, more than half the engineers emerging from the Institutum received practical training working along the Danube or within the Danube valley.[36]

Hindrances nevertheless persisted. Projects on the Danube's tributaries and other waterways, such as the regulation of the Sió and Sárvíz Rivers, lacked coordination and funding, and several officials from Navigation Director Joseph Walcher to Royal Commissioner Károly Sigray to Engineer Böhm eventually ceased work on them.[37] Efforts to drain swampy or water-covered lands, which plagued the flatter, Hungarian territories in the east, encountered resistance from farmers and millers who had grown accustomed to their respective landscapes and feared the consequences of changing them.[38] These wetland commons along the Danube were a vital resource for peasants, not least because of the feudal conditions many still lived under. Worse still, several devastating wars in the eighteenth century forced peasants to billet and feed soldiers. When grain prices spiked during the wars, the landed nobility sought to expand their grain cultivation and increased peasants' corvée labor obligations. Several peasant revolts from 1735 to 1753 along the Tisza, Maros, and Koros Rivers—

tributaries of the Danube—demonstrated peasants' discontent. The danger of mass peasant uprisings prompted monarch Maria Theresa to issue reforms aimed at protecting peasants from aristocratic abuses and expanding peasants' access to arable land—something that was difficult to achieve because of both population growth and the watery makeup of her domains.

Nevertheless, the expanded engineering corps and institutional support did lay the groundwork for more successful and more expansive projects under later emperors from Franz to Franz Joseph. Emperor Franz's early reign (r. 1792–1835), for example, was characterized by large-scale engineering projects that were undertaken even as warfare strained the imperial coffers.[39] In particular, the construction of two major canals reveals the emperor's interest in using new waterways to integrate productive hinterlands to the imperial capital. The first project was the 4 million-florin, 75-mile-long Franz Canal between the Danube and Tisza Rivers through their interfluve region, shortening the southern route between the rivers by 150 miles.[40] Earlier canal designs had failed as engineers were unable to overcome the marshy conditions of the region. By the 1780s, however, chief engineer József Kiss had drained 27,000 square miles of wetlands, an area approximately the size of Ireland, and the reclaimed land enabled 5,000 workers to dig the canal in 1793–1801.[41] After the canal's opening, one of the era's most consequential engineers, István Vedres lauded the project's aims, which "supported the well-being of the fatherland ... advantages from which were expected to benefit not only Hungary but the entire Austrian state."[42] Over 30,000 tons of grain passed through it within its first year, much of that heading to Vienna from Temesvár on the Military Frontier.[43] Converting wetlands to arable land increased land value adjacent to the canal four- to fivefold. The state treasury took in more taxes as a result, living standards increased, and the Bácska region through which the canal passed became one of the most prosperous in Hungary.[44] In 1836, merchants from Arad, a town on one of the Tisza's prominent tributaries, even complimented the strong commercial ties within the empire, thanks to connections between the Danube and Tisza Rivers via the Franz Canal.[45]

The second canal, built in 1797–1803, connected Vienna to Wiener Neustadt about thirty miles south of the capital and provided a critical link to the capital's energy hinterlands. Originally, Maria Theresa had planned to build a canal network connecting the Danube,

Drave, and Save Rivers to the Adriatic, but the project stalled. In the early 1790s, entrepreneurs approached Emperor Franz and convinced him to revive plans in a bid to lower coal transport costs from a nearby mine at Sopron (Ödenburg).[46] Franz approved the project and donated a huge sum from his private wealth to fund its construction.[47] After its opening, the canal transported coal and construction material, including bricks manufactured in factories along its banks. According to one study, "Although transport was the main purpose of the canal, practices of 'secondary' water use soon gained importance. This included energetic use of the canal's slope at the shipping locks as well as water use for industrial and agricultural purposes and forbidden—but nevertheless frequently conducted—practices such as bathing, washing of laundry, watering of animals, bathing horses and extracting fire-fighting water."[48] Even when a new rail line running along the same stretch diminished the canal's commercial importance after the 1860s, the canal continued to support businesses and individual landowners along it for another half century, who leased water rights from the city of Vienna and agitated against filling it in.[49]

Many contemporaries considered both canal projects a success, facilitating movement of food and other resources around the empire and enriching the communities living along them. By linking important industrial and agricultural regions to the Danube, the two projects reveal a similar approach to imperial infrastructure that aimed to grow the empire's economy. Later rail projects like the empire's first horse-drawn railway between Linz and Budweis (1831) followed the same pattern, connecting the Danube's booming salt trade from the Salzkammer region to the important textile industry in Bohemia.[50]

The imperial government played a large role instigating, funding, and managing these hydraulic projects, envisioning their importance for uniting people within a more cohesive imperial economy. Assessing the actions and policies of Habsburg rulers, David Good argues that by the end of the eighteenth century, "the Habsburg Empire was far from being an integrated economic unit. But the evidence presented shows that under the umbrella of Habsburg trade and transportation policy the territories of the sprawling realm were becoming increasingly interdependent."[51] As we will see in chapter 3, this interdependence grew consistently over the next century, built out of the investments that the Habsburgs made in the eighteenth century to boost the empire's cohesion.

An Imperial-Royal Steam Navigation Company

The expansion of waterways and improvements of shipping conditions on the Danube River were the prelude to the nineteenth-century age of steam, when Habsburg rulers mobilized the resources of the state and their personal fortunes to guarantee that steam power existed on the Danube and served an imperial purpose. When Professor Karl Friedrich suggested that steamboats, recently introduced to American, British, and German rivers, "would be of great advantage for the Danube as well" it was more a question of finances than of technical ability to introduce them in the Habsburg Empire.[52] When the state's bankruptcy in 1811 confirmed that the state lacked the funds to directly support such an enterprise, Franz I issued an edict granting exclusive monopoly to any company or individual capable of powering a boat upriver without draft animals.[53] This decree expanded Franz's general patent law from 1810, which he had issued to attract investment and innovation in the empire.[54] Several abortive efforts, such as the launch of the *Franz I*, never translated into a lucrative business. It did not help that press reports detailed steamboat explosions on American rivers, leading some to consider the vessels "Teufelswerk" (devil's work).

The dynasty's support for these early companies finally bore fruit in 1829 when two British engineers, John Andrews and Joseph Prichard, permanently established steam navigation on the Danube. The two men developed their technical skills during their years building steamboats in England, and more important, they gained commercial-financial experience introducing steamboats to the Po River, Lake Geneva, and Lago Maggiore. They founded their joint-stock company, the Danube Steam Navigation Company (Donau-Dampfschiffahrts-Gesellschaft; DDSG), in 1828, and on January 24, 1829, they published a circular to attract financial backing. Among their early investors were the emperor's son Crown Prince Ferdinand, the emperor's brother Archduke Joseph, the emperor's cousin and governor of Galicia Archduke Ferdinand d'Este, the imperial chancellor Prince Metternich, and Hungary's chancellor Count Revitzky. Shortly thereafter, they had sold all two hundred stocks. The emperor granted the company a fifteen-year concession on all steam navigation on the Danube, after which it renamed itself the "Imperial-Royal Privileged Danube Steam Navigation Company." This designation would prove prescient, as the DDSG's burgeoning enterprise became so instrumental to trade and transportation on the empire's waterways that after the 1867 Compromise (Ausgleich), the dynasty, the imperial government, and the Hungarian national government all rewarded the privileged company with favorable financial and regulatory conditions for the rest of the century.

Spearheading efforts to regulate the Danube and introduce steam navigation was the Hungarian aristocrat István Széchenyi (1791–1860). Deeply influenced by his Anglophilic father Ferenc Széchényi, who had traveled to England in 1787–88 and had admired the latest trends in horticulture, architecture, agriculture, and industry, Széchenyi saw steam power as the basis for Hungary's modernization.[55] In his well-known work *On Danube Navigation*, he stated that "steam power and navigation... [have] established an era between past and future, [and] their power will allow even those backward nations to spring forward and catch up to developed ones."[56] For Széchenyi, Hungary was very much the backward nation. The DDSG's founding only underscored some stark economic disparities and financial practices between different parts of the empire. Despite attracting some of the most powerful politicians, bankers, and members of the imperial family, only thirteen out of the company's two hundred issued stocks were bought by men from Hungary, highlighting a dearth of any system of banking or "credit" that would enable investment in such schemes.[57] With the treatise *Hitel* (On Credit), Széchenyi decried situations in which nobles traditionally relied on pedigree and family lands for their income, rather than supporting any banking or financial reforms.

Nevertheless, the Danube influenced many early investment opportunities in Hungary. Following the DDSG's establishment in 1829, stockholders suggested building a shipyard near Vienna to construct and repair boats. Széchenyi pointed out that the river's insurmountable speed upstream from Vienna and its shallowness and sharp bends just downstream from Vienna were reasons to build a shipyard in Hungary instead.[58] His argument won the day, and the company summarily constructed its new shipyard at Óbuda, a suburb north of Buda on the Danube's right bank.

Believing that the benefits of steam power would extend to all people, Széchenyi worked to convince the public to support it. In an 1833 letter to Palatine Joseph, Széchenyi argued that the imperial authorities could dramatically improve the DDSG's fortunes if they guaranteed it wood and coal from state lands and prompted towns along the

Danube to provide planks and gangways for ships to land at.⁵⁹ In 1835, he proposed to the DDSG General Assembly that it demonstrate steam navigation's utility by running the company's newly acquired boat up and down the Danube from Vienna to Semlin and stopping at a few major cities along the way.⁶⁰ The imperial dynasty saw the benefit of such publicity. When Crown Prince Ferdinand (r. 1835–48) became emperor, he visited the DDSG's newest, most luxurious steamer the *Maria Anna*, named for his wife. With Vienna's residents, Ferdinand celebrated the vessel's important task of extending steam navigation upstream from the capital for the first time. When the boat departed for Linz in 1837, communities along the river amassed to catch a glimpse of their first ever steamboat.⁶¹

With a singular but growing steamboat enterprise established on the Danube, the imperial government and the Habsburg dynasty organized the "k.k Central Commission for DDSG Affairs" in 1835 to oversee the company and ensure that it served the public interest well. The following year, Emperor Ferdinand instructed the commission president, his chancellor Metternich, to do all in his power to support the industry: "As the Danube's regulation and navigation is so obviously important for my provinces, it cannot be left to private industry alone, and a thriving result can only be expected if the state administration considers, concerns itself with, and in a sense, guides the whole undertaking, with its various branches and relations, so that collisions or really anything that could hinder its progress are removed."⁶²

The state's paternalistic belief that Danube navigation must remain impartial and imperial faced its first major challenge in 1840. Helping the Hungarian government set up its own steam navigation company, the DDSG's cofounder John Andrews arranged to cede his personal monopoly for Danube steam navigation to the government. When the DDSG protested to the imperial authorities in Vienna, the Court Chancery forced the Hungarians to withdraw their offer. The emperor, however, refused to grant the DDSG's request for exclusive privileges to sail steamboats on the empire's rivers. He did assure its leadership that if the company lowered its freight prices, always checked with the government before raising them, *and* continued to trade along the Lower Danube, then the government would withhold those privileges from any other steam companies.⁶³ This had two strategic benefits. The first incentivized the DDSG to serve the public interest, which advanced the imperial government's interests. The second allowed the imperial government to offer the company its protection from foreign and domestic competition.

In the following decades, foreign challenges influenced the DDSG leadership's decision to forge ever closer ties with the imperial authorities. The company's expansion into the Danube Delta and Black Sea in the early 1830s fell afoul of Russia, which retaliated by allowing the delta's sole navigable channel (under Russian control since 1829) to silt up. After the Ottoman government permitted free navigation to "Austrian" ships in the Black Sea in 1840, it quickly became displeased with the preponderance of ships carrying the Habsburg flag. Declaring the DDSG an unfair competitor, it prohibited its subjects from using the company's ships for transportation, crippling the DDSG's Black Sea business.[64]

The DDSG responded to this antagonism by forming a committee that worked directly with the Habsburg Imperial Court administration (Imperial-Royal Hofkammer Präsidium) to guarantee that the company retain state subsidies. The committee, at the behest of the Präsidium, restructured the entire company in 1844 to place it under more secure management. In 1846, Emperor Ferdinand extended the company's subsidies another forty years, undermining the spirit of all the bilateral agreements that the government had recently signed with Bavaria (1836), Britain (1838), the Ottoman Empire (1839), and the Russian Empire (1840). When Europe's Great Powers internationalized the Danube in 1856 following the Crimean War, ostensibly ending the DDSG's dominance, a new era of state-sponsored subsidies reaffirmed the spirit of Ferdinand's 1846 monopoly agreement and codified the company's control over Danube commerce. In 1857, Emperor Franz Joseph's government provided the company with a subsidy of two million florins in exchange for the company's pledge to offer freight and passenger transport on all navigable waterways in the empire, help maintain that navigability, deliver mail, and transport military personnel.[65] One can contrast this decision with the government's 1854 Rail Concession law divesting the imperial government of owning and operating railways.

Such positioning inoculated the DDSG from some of the worst challenges to its dominance, and its continued vitality likewise protected it from nationalist clamors within the empire. After Hungary regained its autonomy in 1867, for example, some Hungarian parliamentarians and commercial groups petitioned the new government to establish a Hungarian steam navigation company on the Danube. Rejecting these calls, the Royal Hungarian Trade Ministry decided to continue providing the DDSG with subsidies under the original stipulations of the 1846 and 1857 agreements. After seeing the collapse of each national company founded since the 1850s, the Hungarian government perhaps assessed it unlikely that a Hungarian national steam navigation company could compete with the DDSG and pragmatically prioritized support for services that millions of travelers and countless businesses already depended on. After concluding negotiations with the DDSG, the Hungarian prime minister Gyula Andrássy even wrote a letter to Emperor Franz Joseph emphasizing that the Hungarian government had decided on such generous subsidies after weighing the company's benefit for the "communication and interests of the state."

Even after the DDSG leadership felt the state's paternalism was threatening its survival in the late 1870s, it nevertheless continued to rely heavily on the imperial authorities to level the playing field in international affairs.[66] When the Romanian and Russian authorities began subsidizing their national steamboat companies and Romania raised its customs in 1886, the DDSG appealed to the Imperial-Royal Trade Ministry and Hungarian authorities to intervene on its behalf. The Hungarian government complied by subsidizing the DDSG with almost 400,000 florins (over $5.5 million in today's currency) a year to maintain its passenger traffic.[67] In the late 1880s, the DDSG furthermore pleaded with the Imperial-Royal Trade Ministry to speak with the Romanian government, which had begun the practice of allowing only the Romanian national fleet to land at private, well-maintained docks without providing proper landing places for international traffic, contrary to the spirit of the 1878 Berlin Treaty, which had reiterated the Danube policy of free navigation. In 1890, the DDSG wrote to the trade minister and explained that it was losing money every year in its efforts to maintain the Danube's navigability, keep up public harbors, and fight against rail competition and Romanian tolls. It concluded by stating that the company served the best interests of everyone in the empire, "in Austria and Hungary," and it was important for the state's *prestige* that it retained shipping dominance on the Danube, especially vis-à-vis its Lower Danube rivals.[68]

Despite challenges beyond the empire's borders, at the end of the century the DDSG's dominance in the empire was total. According to the Central Statistical Office in Budapest, the company still transported 70 percent of the river freight and, excluding ferry traffic, 98 percent of river passengers in Hungary in 1889.[69] While the Hungarian government finally acquiesced to domestic political pressure and established a national company to challenge the DDSG's hegemony in 1894, it did so on the DDSG's terms. During negotiations, the DDSG secured the rights to continue operating on the Middle and Lower Danube and mandated that the new company share shipping routes to Romania and Serbia and around Budapest. Outpacing its new Hungarian counterpart, the Hungarian Sea and River Stock Company (MTFR), the DDSG's freight business continued to expand in Hungary, averaging more than two and half times the number of passengers transported on the Danube in Hungary, and on the "Hungarian" river, the Tisza, its passenger traffic was on average more than 10 percent greater than the MTFR's.[70] Although it is not always clear *who* traveled with each company, it is possible that passengers chose the DDSG due to the company's convenience, competitive pricing, and perhaps even out of "brand loyalty" over pure, nationalist calculations to support a "Hungarian" company.

A brief history of the Danube Steam Navigation Company provides valuable insight into the mechanics of collaboration between public and private enterprises, particularly one that advanced so many of the Habsburgs' agendas for empire-wide travel, navigability, and connections along the Danube. Supporting technical advancements through their own personal fortunes and intervening in the economy to bolster private innovation along the Danube, Habsburg monarchs took great pains to keep steam navigation firmly oriented toward desirable imperial endeavors. Imperial and national governments interpreted the impulses emanating from the rulers and drew up formalized state sponsorships to protect the Danube Steam Navigation Company from financial difficulties and safeguard it against international and domestic competition. The Habsburgs were sincere in their belief that such actions would benefit the empire through greater domestic trade and bring together citizens via more comprehensive and cohesive transportation options.

Public Consumption on the Imperial Danube

Having invested in the "improvement" of rivers across the empire as well as the expansion of steam navigation, the Habsburgs' approach to environmental engineering, like their governance more generally, evolved in the nineteenth century, as they adjusted to an increasingly democratized, more politically and socially engaged public. The "public sphere" emerged as a metaphorical place for political dialogue and social discussion, yet manifested itself in real, physical places where people met to exchange and express ideas. With steamboat travel, newly regulated river stretches, and massive infrastructure projects from bridges to quays along the river, new spaces emerged where the imperial family was able to include the populations in curated celebrations honoring the dynasty's role in transforming the Danube to facilitate commerce and travel and to promote safety and recreation. Daniel L. Unowsky has argued in *The Pomp and Politics of Patriotism: Imperial Celebrations in Habsburg Austria, 1848–1916* that imperial celebrations in the mid- to late nineteenth century mirrored medieval baroque courtly rituals and ceremonies meant to reinforce people's devotion and loyalty to their monarchs. This was a critical part of shoring up the legitimacy of monarchical governance, requiring more active outreach and engagement between the dynasty and its people. In some instances, the imperial court worked with provincial and municipal authorities to devise elaborate exhibitions that brought together the dynasty and the public to showcase their projects and initiatives designed to win their loyalty and "patriotism."[71]

Along the Danube, members of the imperial family demonstrated their care for their citizens' well-being in a multitude of ways. Their physical presence and monetary donations comforted victims of floods, and they publicly praised and awarded engineers, officials, companies, and communities that demonstratively improved the Danube's utility for the public. In doing so, they strove to retain and strengthen the dynasty's legitimacy as a source of benevolence for the common good in a time of international competition, internal strife, and revolutionary environmental change.

Imperial-Royal Family in the Flood

A paper from Pécs, Hungary, published the following anecdote about Archduke Joseph during an unidentified flood on the Maros River in Transylvania:

> When the flooded Maros threatened to flood Arad and the whole region, Archduke Joseph was there at Kis-Jenő, where the danger was even more threatening due to Kőrös (River). One evening, when the water of the river had reached the top of the levee, the Archduke anxiously asked the vice deputy [*alispán*], if he had taken precautions at night, were they guarding the dam sufficiently? "I have taken all measures necessary,"

Creating the Imperial Danube

Figure 1.3. *Prince Anton Victor Visits Flood-Struck Viennese*, 1830. Johann Nepomuk Höchle © Wien Museum.

replied the *alispán*; "Your Majesty can be at peace." And with that, the *alispán* fell asleep peacefully. But the Archduke was kept awake by concern. It was still a dark night as he hurried out alone to the deserted levee, where he found no soul but the flood, which had already made its way through and was rushing out the levee. Archduke Joseph roused the people to their feet and then to the levee, where he himself carried earth-filled sacks to the gap, wading in water, until he finally filled it with earth, preventing its destruction. When he had finished, he returned to the nightly accommodations. Opening the *alispán*'s bedroom door, he stood at attention, saluted, grinned, and said, "I humbly report, honorable *alispán*, that we have filled the gap, the danger is gone, now you can sleep peacefully!"[72]

In the early nineteenth century before large-scale flood protective measures were in place, floods were a regular part of life along the Danube and other rivers, which devastated the empire's riparian communities. The dynastic family's presence in flood-stricken areas, sometimes during the event itself, provided reassurance to residents that had in many instances lost homes, property, or, at worst, family members. A particularly popular work written by Dr. Franz Sartori, *Vienna's Days of Danger and the Saviors from the Peril,* lionized Emperor Franz and his whole family's work to alleviate the suffering caused by the massive 1830 flood (fig. 1.3).[73] In the early nineteenth century, cities, provinces, and even the imperial government lacked formal mechanisms to help these people. Like Archduke Johann stacking earthen sacks in the breached levee, the imperial family stepped in to fill the void. Besides organizing relief efforts and mandating public funds to compensate those who lost property, the dynastic family generously funded relief aid. After the 1830 flood at Vienna, Franz donated 40,000 florins and his wife donated 12,000 florins to a newly formed commission for immediate distribution to the afflicted populace.[74] After the 1838 flood in Hungary, Emperor Ferdinand and his wife, as well as other Habsburg family members donated a total of 117,000 florins to the victims.[75] These were immense sums: an agricultural worker made on average 6 florins a month in 1840.[76]

Even after the authorities established more institutionalized structures to aid the populace during and after floods, the dynasty remained a critical source of comfort and support. According to Daniel Unowsky, "Franz Joseph's charitable contributions, publicized in the official meeting, merged his personal generosity with the commitment of the state to address the economic difficulties facing his subjects. . . . On every imperial celebration and in response to economic downturns and natural catastrophes, Franz Joseph made large and very public donations to the needy and to institutions supporting the public welfare."[77]

Paintings and lithographs immortalized the Habsburgs' magnanimity, and contemporary papers and pamphlets widely published these images for the public's consumption.[78] Franz Joseph, Elisabeth, and Rudolf's visits to flood-stricken sites across the empire in 1862, 1876, 1879, and 1884 appeared in the papers, as did the records of their generous giving (fig. 1.4). After coaxing provinces and his own imperial bureaucracy to spend more on flood prevention measures, Franz Joseph visited the sites of new embankments to praise progress and

Creating the Imperial Danube

Figure 1.4. *Franz Joseph I During the 1862 Flood.* August von Panttenkofen (1822–1889).

remember victims of previous floods.[79] The newspapers lauded the dynastic members' appearance and documented the widespread appreciation for their support.

Heidi Hakkarainen has argued that during times of flooding and crisis, newspapers and the Habsburg family both reiterated the same tropes emphasizing the dynasty's ability to reestablish order and help citizens recuperate from the flood's aftermath. Hakkarainen claims that portrayals of Franz Joseph and other Habsburg figures rescuing the people perpetuated a narrative of masculinity and dominance that completely neglected women's contributions to rescue and recovery efforts.[80] Such criticism fails to note that newspapers frequently celebrated and distinguished female contributions from the dynastic family, such as from Emperor Franz's wives as well as from Franz Joseph's wife, Empress Elisabeth, who insisted at the time of her and Franz Joseph's twenty-fifth wedding anniversary in 1879 that people donate to Szeged flood relief funds rather than gift anything to the imperial couple. Furthermore, a local actress named Therese Krones anecdotally braved the cold, wet weather to aid rescue efforts in 1830. She contracted a lung infection during her work, which later proved fatal.[81] After her untimely death, not only was she widely mourned but on the hundred-year anniversary of her birth, her home district in Vienna put on a Therese Krones Day to memorialize her time on the stage and her valiant work during the Danube's flooding.[82] Floods were thus among many occasions for the imperial family to actively and compassionately help their people, donating to relief efforts, visiting afflicted communities, and mourning and comforting those in need.

Sites of Celebration, Locations of Loyalty

Floods were extraordinary events that brought the dynasty to the Danube in time of need, but urban renewal works in the nineteenth century provided a more joyous occasion for the Habsburg family to celebrate the empire's rivers. Eighteenth-century industrial growth and agricultural innovation triggered sustained population growth in the empire and around the world, bringing with it surging urbanization.[83] Cities were no longer physically large enough to house rapidly growing urban populations, and engineers thus widened boulevards, sent buildings upward, reclaimed floodplains, created new public transportation, and crisscrossed waterways with rail and pedestrian bridges in an effort to expand and connect aesthetic and utilitarian elements in the city. New constructs such as bridges, quays, and embankments fostered connections between citizens, underpinned municipal growth, and embodied dynastic largesse and prestige throughout the nineteenth century.

Presiding over opening ceremonies, members of the Habsburg family celebrated the empire's newest civic engineering and architectural constructs, joining in on a custom that was quickly gaining in popularity for heads of state around the world, monarchical, republican, democratic, and despotic. These events provided local populations with an occasion to see their monarchs in person, an occurrence that had slipped away as empires outpaced the ability of leaders to travel regularly through them. European monarchs took great care to cultivate these ceremonial events. Invited by the Canadian Parliament in 1859 to attend the opening of the Victoria Bridge, Queen Victoria sent in her stead her eighteen-year-old son Prince Edward, whose public appearances set up a successful model for future royal visits.[84] The indefatigable Wilhelm II of Germany inaugurated countless dams and bridges throughout his empire, seeing them as symbols of the empire's industrial might and technical prowess.[85] Other contemporary rulers, from Tsar Nicholas II of Russia to Napoleon III of France to Abdülhamid II of the Ottoman Empire, all likewise maintained a relatively full schedule of public events inspecting and championing engineering works that commemorated leaders and significant events in the state's development.[86] Completed sites reified civic or mythical connections with figures and moments significant to both a state's history and its contemporary development.

For the Habsburgs, as the century progressed, new riverine infrastructure became more common and symbolically important as the dynastic state sought to legitimate and consolidate popular support for itself. Emperor Franz Joseph (r. 1848–1916) actively and frequently participated in these imperial displays along the Danube River, because they showcased his legitimacy and concern for his people's well-being at a time in which mass politics and ideologies like socialism, nationalism, and liberalism provided appealing alternatives to semi-constitutional monarchies. Unable to rely on ethnolinguistic or nationalist rhetoric to unite his diverse citizenry or guarantee support for the dynastic-imperial order, Franz Joseph instead dramatically combined massive public works along the Danube with large, public ceremonies to strengthen the popular association between monarchy and modernity.

In one of earliest and grandest gestures, a string of celebrations in 1854 greeted the arrival of a special steamboat arriving in the Habsburg Empire from Bavaria. Aboard the newly launched steamboat *Franz Joseph* was Duchess Elisabeth Amalie Eugenie, the future bride of Franz Joseph. As Elisabeth's boat sailed into Habsburg territory, the towns and communities, it passed buildings near the waterfront bedecked with flags and garlands. Residents crowded the banks to wave at their future empress and bands greeted her with music. Her

Figure 1.5. Sissi's Arrival in Linz, Austria. Joseph Edlbacher, *Der Abschied* [The Farewell], 1854. Photo: Nordico Stadtmuseum Linz.

first stop on Austrian soil was Upper Austria's provincial capital Linz, where cannon salutes and church bells pealed out across the river as her steamboat docked (fig. 1.5). Unbeknownst to Elisabeth, her future husband Franz Joseph had also boarded his own steamboat in Vienna to ply upstream and surprise her when she arrived in her new homeland. After her reception in Linz, she and Franz Joseph continued to Vienna, where all court dignitaries, religious figures and ministers, civil servants, and aristocrats welcomed her with all the official pomp of the court.[87] The *Wiener Zeitung* covered the momentous occasion, indicating the joy and enthusiasm that the young woman inspired throughout her journey along the river and throughout the empire.[88]

Not unlike the earlier voyage of *Franz I* in 1823, Elisabeth's journey represented a successful unification of two elements: the imperial state's choreographed displays to foster popular loyalty to the dynasty *and* the Danube's backdrop as a site of state-sponsored innovation and modernity. In contrast to earlier retinue traveling around the empire, which featured horse-drawn carriages and laboriously slow progress along muddy roads, Elisabeth's steamboat arriving on the Danube in the early 1850s passed extensive new embankment construction works, new landing places, and new urban designs. Steamboats themselves were becoming more modern, transitioning from wood to iron-hulled ships after 1839. The dynasty's use of steamboats and a dynamic environmental backdrop for such an important ceremony indicated a continued willingness to embrace and employ technology as a tool to underpin its legitimacy.

The arrival of the young Elisabeth from Bavaria in 1854 coincided with a decade filled with challenges to the Habsburg Empire and especially to the dynasty. The eighteen-year-old Franz Joseph had ascended to the throne during the chaotic 1848–49 uprisings, a violent outpouring of socioeconomic discontent and nationalist uprisings in Hungary, Bohemia, and Italy. Once the last of the uprisings were repressed in 1849, Franz Joseph reprised Joseph II's absolutist governing style, abrogating provincial diets and centralizing power in the hands of his imperial bureaucracy and government in Vienna. To dispel liberals' displeasure with the neoabsolutist regime, he instructed his government to implement social and economic reform programs and invest heavily in domestic infrastructure works. Foreign policy, however, was a disaster. He remained neutral during the Crimean War (1853–56) facing European-wide opprobrium. Nicholas I was especially incensed by Franz Joseph's position given that the tsar had sent Russian troops to help repress rebellious Hungarians in 1849. A military defeat in 1859 against united French and Piedmontese forces lost Franz Joseph one of his wealthiest provinces, Lombardy. The resulting state bankruptcy forced him to reinstate limited constitutional governance to secure loans from wealthy middle-class bankers. Franz Joseph's personal involvement in the military meant that the disastrous outcome reflected poorly on his leadership.[89]

Against this backdrop of political and military setbacks, the young emperor and his wife focused on the promotion of Danube engineering projects designed to improve municipal life, intraimperial travel, and commerce for the empire's inhabitants. Hoping that such works would resonate with the people and shore up support for the young emperor, Franz Joseph and Elisabeth undertook a series of displays that included celebrating new infrastructure, honoring and ennobling technical workers, and invoking the importance of such works for local development and imperial unity.

Creating the Imperial Danube

The opening of new bridges, for example, became a constant source of celebration and ceremony, and provided local populations throughout the empire a grand occasion to celebrate the imperial pair. Elisabeth's wedding procession into Vienna officially began by crossing a newly constructed bridge over the Vienna River, which subsequently carried the name "Elisabeth Bridge."[90] Residents of Děčín named a newly constructed suspension bridge, the largest in Bohemia, after the empress in 1855, and the following year, the imperial couple visited Carinthia's provincial capital Klagenfurt to open a new bridge. During the huge festival, the imperial pair proceeded through town, surrounded by the imperial, provincial, and Bavarian colors and crowded on all sides by its inhabitants. After reaching the bridge, meeting its engineer, and crossing the flower-strewn path, the empress "graciously permitted it to be called 'Elisabeth Bridge' in remembrance of the day's celebrations" to the joy of the city's inhabitants.[91] These displays provided local inhabitants with a venue to receive and honor Empress Elisabeth, who was widely adored in the empire because of her relatively humble upbringing in a minor branch of the Bavarian royal family.

Empire-wide flooding in 1862 accelerated plans to regulate the Danube and led to renewed initiatives, sanctioned and celebrated by Franz Joseph, to protect citizens from the capricious elements of the environment. Plans at Vienna and Budapest called for straightening and deepening these stretches to prevent ice formation, raising embankments to protect from flooding, and constructing quays, steamboat landing places, factories, and warehouses to facilitate trade and traffic along the newly regulated river. The Viennese municipal authorities set up the Donau-Regulirungs-Commission in 1866 and operations began the following year, funded jointly by imperial, provincial, and municipal legislatures. Following the 1867 Ausgleich, which devolved control of Hungary's internal affairs to its National Diet, lawmakers likewise passed legislation in 1870 setting up the mechanisms to fund and direct the Danube's regulation. The organizing authority, the Capital Public Works Advisory Council (Fővárosi Közmunkák Tanácsa) was modeled on London's Metropolitan Board. Besides the Danube's regulation, the council designed new public transport networks around Buda and Pest and oversaw the cities' demolition and reconstruction, which was similar to the Haussmannization of Paris.

Imperial and national authorities greenlighted funds for these municipal projects because frequent flooding and hitherto short-lived

Creating the Imperial Danube

Figure 1.6. Crowds Cheer Franz Joseph's Steamboat During the Opening of the Regulated Viennese Danube, 1875 © Brandstätter Verlag.

regulation works made it clear to leaders that the Danube's regulation was a task far greater and more important than any single location could hope to fund or direct. Such an investment promised to benefit all. To the cheers of thousands of Viennese and in the presence of notables, archdukes, the chancellor, and members of the communal and provincial diets, Franz Joseph spoke at the groundbreaking ceremony for the Viennese Danube Regulation, pledging that the new riverbed would have "blessed repercussions" not only for the imperial residence at Vienna but for the entire monarchy.[92]

When the work was finished in May 1875, Vienna's population amassed along the banks and bridges to witness the river's release into its new bed. Franz Joseph appeared and acknowledged the crowds, as his interior minister thanked him for his support for this project. To the sound of twenty cannons, Franz Joseph boarded a steamboat to sail upstream to Nussdorf, admiring the new quays and facilities on the riverbanks and waving to the throngs of people he passed along the way (fig. 1.6).[93]

The Habsburgs' names became synonymous with the modern infrastructure emerging along the Danube. Newly constructed walk-

51

ways, bridges, quays, and other sites bore the names of members of the dynastic family. Before Vienna hosted the 1873 World's Fair, several bridges were built or rebuilt to facilitate the anticipated foot and tram traffic. Vienna's first suspension bridge, which had been opened in 1825 and named for Archduchess Sophie (Sophiebrücke), was replaced with a massive iron suspension bridge capable of handling larger cart and tram traffic across the Danube Canal from the inner district to the World's Fair grounds at Prater.[94] The Kaiser Joseph Bridge was built and honored the monarch who had opened the Prater to the public in the eighteenth century.[95] Several other stone bridges were constructed to span the Danube, increasing road and rail connections between the empire's northern and southern provinces.

New infrastructure also facilitated commercial connections between cities along the Danube. After regulation works were complete, contemporaries praised the new quays for their modernity and connectivity. In Vienna, six miles of quays flanked by warehouses, rails, factories, and living complexes ensured an industrial and commercial connection to the Danube.[96] Observers in Budapest claimed that the 12.5 million florin price for regulating this segment of the Danube had improved navigation dramatically, and the two and half miles of quays likewise enjoyed lively river traffic.[97] The Franz Joseph Promenade (Ferenc József sétany) and Rudolf Quay (Rudolf rakpart) in Budapest and the Franz Joseph Quay (Franz-Josef-Kai) in Vienna provided the residents with space for fish and fruit markets, tree-lined promenades, and modern streetcar lines. These quays also had steamboat stations, which enabled passengers to either conveniently cross the Danube, as millions did annually between Buda and Pest, or to take ships up and down the river, which millions more did.[98]

Decades later, infrastructure works were still the site of dynastic displays. The new Franz Joseph Bridge opened in 1896 as part of Hungary's Millennial Exhibition and millions of people from around the empire traveled to the capital throughout the six months of celebrations.[99] Thousands attended the bridge's inauguration in October, including Franz Joseph I and Elisabeth, who had traveled to the capital to participate in the unveiling ceremony. Because the city had named the bridge for him, Franz Joseph proudly hammered his initials "F.J.I." in silver rivets on the bridgehead. Local papers offered positive coverage of the event. The *Pesti Napló* discussed the bridge's importance and history for the city, and the *Pester Lloyd* praised the monarch's atten-

Figure 1.7. Franz Joseph Crossing the Newly Opened Franz Joseph Bridge in Budapest, October 1896. *Source: Vasárnapi Ujság*, October 4, 1896.

dance at the bridge's opening as an important part of the millennial festivities.¹⁰⁰ The front page of the *Vasárnapi Ujság* featured a prominent picture of the crowds following Franz Joseph on foot over the bridge (fig. 1.7).¹⁰¹ Postcards of the bridge multiplied after its opening. While the new bridge played into the Hungarian national situation—particularly the pride that Hungarian architects, materials, and technicians were used to construct it—Franz Joseph's presence and positive reception at the new Danube arrangements demonstrated Hungarians' renewed reverence for the dynasty since the 1867 coronation.

Such imperial-royal displays along the newly engineered riverfront also took place in cities and towns outside the dual capitals. Expansive regulation works in the 1880s and 1890s, heavy steamboat traffic on the Danube, and readily accessible iron and steel encouraged communities to replace pontoon bridges, flying ferries, and other more traditional passages across the river with modern bridges, which were less vulnerable to flooding and ice flows and would not hinder river navigation. A series of new bridges opened during this period, which local communities also named after the imperial family. In 1883, Neusatz/Újvidék (modern-day Novi Sad, Serbia) opened its new Franz Joseph Bridge. In 1890, Franz Joseph traveled to Pozsony/Pressburg dressed in a Hussar's uniform and, along with masses of local people, processed through decorated streets and victory arches to the newly constructed Franz Joseph Bridge, from which he exclaimed in familiar fashion that the new bridge would serve "the interests of Pressburg, the region, and the entire monarchy." He also expressed his joy that the bridge represented the "triumph of technology" that would stand as a testament for later generations.¹⁰² The event was covered in papers throughout Hungary.¹⁰³ Hungary's preeminent engineer János Feketeházy designed the Elisabeth Bridge to connect Komárom and Új-Szőny, and Franz Joseph attended its opening in 1892 with the Hungarian prime minister.¹⁰⁴ Residents also named the small island near the bridge after Queen Elisabeth, "in remembrance of the fact, that Her Majesty stepped on Hungarian ground [there] for the first time on May 4, 1857."¹⁰⁵ Feketeházy later designed a bridge in Esztergom, opened in 1895 to great celebration, which Franz Joseph permitted to be named for his youngest daughter Maria Valeria, to whom Elisabeth had given birth in Hungary.¹⁰⁶

Such vast new construction projects, riverine infrastructure, and celebrations further cemented the association between monarchy and modernity and demonstrated the population's deep-seated affection for the monarch. Symbols of this connection abounded in both grand and trivial ways and through both joy and sorrow for the imperial family. In 1898, booksellers, merchants, and others sold literature, trinkets, and numerous objects to commemorate Franz Joseph's fiftieth jubilee. One official publication described the young couple's prewedding steamboat trips on the Danube and the mass festivities that accompanied Elisabeth's arrival into the empire.[107] As part of the celebrations, the Danube Regulation Commission set up a huge exhibit to educate people about the dynastic family's support for the hydraulic projects taking place on the Danube.[108] Over 2.5 million people visited the Jubilee Exhibit grounds.[109] But as celebrations were underway, an Italian anarchist assassinated Elisabeth while she was visiting Geneva. Elisabeth's death led to empire-wide mourning and countless memorials dedicated to her, including sites connected to the Danube.

In February, before Elisabeth's assassination, Franz Joseph had traveled to Budapest to lay the ground stone for a new bridge that would be named "Eskütéri híd" (Eskü Square Bridge).[110] Two months after Elisabeth's death, Franz Joseph consented to have the bridge renamed "Elisabeth Bridge," in his wife's honor.[111] When the bridge's construction was completed in 1903, local Budapest papers advertised that the king would personally come to Budapest to open it.[112] At the same time, towns and communities across the Hungarian lands petitioned to name parks after the queen because of her love of nature.[113] On the one-year anniversary of Elisabeth's death, the mayor of Linz gave an impassionate speech in her memory, after which the city council unanimously voted to name their newly regulated Danube stretch "Empress Elisabeth Quay" after the woman, who had "taken her first step in her new home at this spot."[114]

This devotion to Elisabeth lingered even after the empire's collapse in 1918. Budapest's residents erected a statue to her in 1932, which overlooked the Elisabeth Bridge—the only dynastically inspired bridge name that the Hungarians kept in the capital in the postwar period. In 1926, the residents of Esztergom on the Danube likewise commissioned a statue of the queen.[115] The river's engineering had provided imperial and local authorities a means to modernize transportation and trade and to protect the material well-being of the population, while fostering devotion to the dynasty. In such instances after Elisabeth's death, the built environment became a site of remembrance, which the residents dedicated to a beloved monarch.

Strengthening the Dynastic and Public Well-Being Along the Danube

From the mid-eighteenth century onward, the Habsburg dynasty took an active role in transforming the empire in its pursuit of a stronger, more unitary state. Passing political and economic reforms and sanctioning engineering works promised to strengthen the imperial state's financial, political, and military foundations as well as its social and economic cohesion. These projects also revealed efforts by the dynastic-imperial-bureaucratic state to legitimate itself by making the care of its citizens its raison d'être. James Shedel calls the Habsburgs' new approach to governing the "eudaemonic principle," and the Danube became a central site where the dynasty applied this newfound ideal.[116]

Engineering the Danube, whether constructing canals, commissioning bridges, supporting steam navigation, regulating unruly stretches, or erecting flood embankments, sought to project the dynasty's vision for a nonnational, imperial Danube, which served the common good. These projects were intended to enhance the Danube's safety and utility for citizens in the empire, with the intention that it would bolster their identification with and loyalty to the imperial state. These works also served to unite far-flung provinces and disparate populations. To pursue these goals, the dynastic family assiduously supported innovations that fundamentally changed the relationship between society and nature.

To demonstrate the practical and positive benefits conferred by these works, the dynasty participated in popular ceremonies that celebrated local works surrounded by local populations. These ceremonies provided a visual means to reify associations between the Danube's improvement and the dynasty, while fortifying the river's place at the heart of the empire. These ceremonies also provided local actors and populations with a space to display their devotion to the dynasty, whether through the naming of infrastructure after members of the dynastic family or appearing and cheering at their events. All this provided a sense that the dynasty retained its legitimacy as a traditional source of authority, prestige, and protection for the populace while offering people modern experiences and enhanced opportunities and livelihoods along the Danube River.

Chapter 2

THE DANUBE AS LIFE ARTERY

In October 1862, the geographer Vinzenz Klun stood before the Imperial-Royal Geographical Society to unveil maps of the Danube and Tisza Rivers that the Imperial-Royal Statistical Office had spent over a decade producing. The scale and detail were immense. The map of the Danube was over one hundred feet long and made up of ninety individual prints.[1] At a glance, the viewer could discern every rapid, meander, and island in the river from Passau to Orsova. Symbols along the river indicated factors that affected navigation and flood conditions. Beginning his speech, Klun deployed a common metaphor of the rivers as "arteries" through which "pulsated the life traffic of the people." Drawing a vivid picture of the Danube's specific significance and centrality to the empire, he declared, "It is the main artery for all commercial life in our fatherland; its course passes through the heart of the imperial state, as well as through flourishing, rich Hungary, and the southern regions rich in agricultural products. The navigable rivers of most of the crown lands connect to this great waterway.... The full virility of the glorious river, the mighty Middle Danube, belongs entirely to Austria, and nature makes of Austria a coherent, dominant Danube Empire [Donaureich]."[2]

Coherence, in this case, was linked as much to the Danube's geography as to its commercial value. Drawing different tributaries and crownlands together, it served to unify complementary industrial and agricultural regions. Such ideas were not born of mere theoretical study of the empire's geography but of Klun's practical experience working in the Laibach Chamber of Commerce, teaching at the Trade Academy in Vienna, and later as a representative to the Imperial Diet (Reichsrat) after 1867. Those familiar with Klun's lectures at the university or who had read his *General Geography with Particular Consideration of the Austrian Empire*, published the year before in 1861, would have heard in his speech variations on a similar theme—namely, that the empire's physical geography, including its rich waterways, was well-suited to address the needs of those living within it.[3]

Many contemporary scholars in the empire held similar views about the empire's favorable geography. As Deborah Coen has written, naturalists and, later, scientists traipsed across the Habsburg Empire to understand how disparate, diverse pieces of information about the empire's geography and climate fit into a "coherent," imperial whole. Some engaged in measuring minute details of local rainfall, soil moisture, temperature, wind speed, or air pressure, while others busied themselves illustrating local biota from vast sweeping vistas of the Hungarian puszta (steppe) to intimate marshland vegetation tableaus or sketching varying geological and alpine formations. They often traveled (or were posted) to different parts of the empire and were confronted by data that varied across time and space. Striving to discern meaning from this plethora of information, these scholars relied on sharing and comparing profuse and varied observations to synthesize details about the empire.[4] Like his itinerant colleagues, Klun's work in trade, journalism, education, and politics, and his activities as a member of literary and hiking clubs, took him from his home province of Carniola to Dalmatia and eventually to Vienna, allowing him to formulate broader opinions about his homeland.

Ending his speech for the Geographical Society, Klun asserted that ongoing regulation works along the empire's rivers only enhanced the benefits of these efforts to society. "With true satisfaction, even with justifiable pride, we can point to the patriotic undertakings of the Danube and Tisza regulations. Such eloquent monuments of energy for raising the material well-being of a land ... stand unshakable, imperishable."[5] These were fine words, but what did they mean from a practical point of view? How did those "monuments of energy" unfold, and in what ways did people envision that they would raise the material well-being of the empire's population?

This chapter reveals that regulation works and other hydraulic engineering projects loomed large in economic and social questions for people living in the Habsburg Empire. From a historical perspective, modifications to the Danube had typically been undertaken by local communities or provinces with little ongoing oversight or support from central authorities. In the nineteenth century, however, river regulation became tied up in questions of state-building, nationalism, imperial identity, and democratizing processes. This work eventually involved dozens of companies and ministries and thousands of workers throughout the empire who spent decades digging new courses, erecting levees, and wrestling the Danube and many other rivers like it into a single bed. Their work was championed by government officials, landowners, merchants, academics, and others who wrote about sluggish trade, waterlogged landscapes, and devastating floods as the natural and adverse consequences of the unregulated river. These efforts were coordinated by a growing cadre of public works officials who, like Klun, took for granted that engineering the Danube would bring material prosperity. Civil society and "peripheral" regions, for their part, assumed rhetoric like that of the imperial-national authorities to attract development and promote local interests.

Through all of this, the Danube and its environs remained no mere passive feature for the monarchy's cadre of engineers to modify. As actors attempted to regulate, straighten, and control the river, they eventually realized that the Danube's "behavior" (if we may personify hydrological circumstances influenced by geological conditions, human practices, and meteorological occurrences) also dictated the framework for their success or failure. As a result, the Danube guided governmental investment indiscriminately across territorial and lingual enclaves, as actors realized the need for the holistic management of regulation. Even as particularistic sentiments divided the government and occasionally led to political logjams, the Danube remained a space where cooperation and a joint commitment to advancing material benefits for citizens maintained a slow but continual progression of hydraulic works and investments until the empire's end.

Challenges and Opportunities of the Unregulated Danube

A few decades before Klun's speech, in the early years of the nineteenth century, the Danube was still a highly unregulated river with navigational obstacles and an ever-meandering course. One 1813 guide cautioned, "It has mountainous banks and, in many places, sharp rocks in the shipping lane," despite work done by the Navigation Directorate in the 1770s and 1780s to blast them.[6] Large, rocky islands in the river, such as "projecting rocks" at the *Strudel*, affected the Danube's current, creating eddies and other perilous passages, such as the *Wirbel*, which was typically confined to small, lightweight crafts.[7] The river's speed on its upper reaches eroded riverbanks, transporting and depositing sediments in slower segments of the river downstream. This natural and perpetual process formed meanders, gravel bars, and new, branching arms.[8] Gravel accumulated around the frames of sunken ships and tree trunks, which had been dislodged in prior flooding, forming new islands in the river and complicating navigation.

Passing from narrow valleys to wider floodplains, the river broadened unencumbered, growing slower, shallower, and swampier. A geographical work in 1829 described how "the river expands anew in its course through the Hungarian plains, forms large islands, and passes through a country of which the inclination is not more than twenty inches in the league. Its banks are covered with marshes in the southern part of Pest, in the districts of Bacs and Tolna towards the confluence of the Drava."[9] Marshes formed wherever water drained slowly or not at all and were undesirable from a navigational perspective, as they were a telltale sign of shallow water and amorphous riverbanks.

For over a century, government policies in Hungary had prioritized regulating rivers and draining floodplains to make land near the river habitable and to encourage settlement. Officials targeted in particular the empire's depopulated southern and eastern territories, which had been under Ottoman occupation until the late seventeenth and early eighteenth centuries. Field Marshall Claudius Florimund Mercy, the governor of the Temesvár Banat (bordering the Ottoman Empire), built canals to regulate and drain land around the Béga and Temes Rivers in the 1730s. The military engineer Maximilian Fremaut continued Mercy's work draining these borderland regions to repopulate them with settlers in the late 1750s and 1760s.[10] József Kiss made progress draining marshes in the Délvidék region along the Danube near the Drava, Danube, and Tisza confluences before construction of the Franz Canal in the last decade of the eighteenth century. Statistics indicate that population growth in 1790–1855 was greatest in areas where arable land was increasing, mainly in counties within river floodplains where wetlands were being converted to fields.[11]

Drainage projects brought with them social and ecological dislocations. Peasants used marshes as supplementary pastures for horses, pigs, cattle, and sheep. Pigs rooted around in marshes eating worms, bird eggs, or vegetation, and it was not uncommon for pasture-fed cattle to be driven to wetlands for the winter months to nourish themselves on tubers, roots, and shoots.[12] For peasants without access to arable land, raising herds with the help of wetland foraging was a valuable source of income.[13] Wetter parts of the floodplain offered fishing and gathering opportunities, drier patches above the flood line were available for beekeeping, and mixed landscapes near the river were ideal for hunting.[14] When a contemporary work claimed that in Hungary "the inhabitants [were] anxious to diminish the number of marshes," it most certainly did not consider the livelihood of those who relied heavily albeit informally on them.[15] Despite or because of this informal usage, drained marshes and arable land were preferred landscapes, enabling nobles to profit from rising grain prices, and governments to secure more food for the growing population and revenue from more easily taxable agricultural products.[16]

While such drainage measures had accompanied regulation for centuries, the need to increase rivers' navigability gained urgency with steamboat trials in the 1810s and 1820s. Introducing steam navigation to the Danube and other major rivers promised to remake the empire's extensive river network, turning it into a significant piece of imperial infrastructure that knit together internal and international trade and travel. In 1819, the director of the Imperial-Royal Polytechnic Institute in Vienna, Johann Joseph Prechtl argued that France, Britain, and above all America were using their waterways to greater economic effect and that "connections have significantly increased thanks to the economy and rapidity of their steam navigation."[17] Emperor Franz commissioned two surveys of the empire's rivers in 1817–19 and in 1823–38 to speed along plans for their regulation.

The Danube Steam Navigation Company's successful expansion after 1829 and a ruinous flood in 1830 both provided a short-term boost to regulation spending. In 1831, hydraulic engineering expenses in the

Austrian, Bohemian, and Italian lands amounted to over two million florins, equivalent to one-tenth of all government-issued currency that year.[18] Hungary's Lieutenancy Council mandated the regulation of the Danube's main bed between Pozsony/Pressburg and Gútor in 1831, and by 1837 had promised a total of two million florins to continue work as far as Védek, a town halfway between Pozsony/Pressburg and Pest-Buda. In 1833, István Széchenyi was appointed royal commissioner for the Danube by Emperor Franz's brother Joseph, the popular palatine of Hungary. Széchenyi used his own funds to purchase dredgers from England to support regulation work. The following year he claimed to have fifty letters attesting to the public's support for the Danube's regulation. He also acknowledged that many feared that river trade would irreparably alter their way of life. Seeking to alleviate these concerns, Széchenyi argued that clearing the Danube's riverbeds was a "spiritual victory" and "patriotic undertaking" that would help unite the people.[19] An 1836 commercial guide took up this mantra, arguing that the imperial government's growing expenditures on regulation "advance[d] the empire's unity" and made the state "ever more a whole," while expressing hope that one day "the bright colors of the long faded national differences would disappear for the health of culture, civilization, and general well-being."[20] By January 1840, Austrian and Hungarian newspapers reported that Emperor Ferdinand I was so convinced that the Danube's regulation would benefit "national well-being" that he encouraged the raising of salt prices in Hungary to ensure that engineers had the resources they needed to be successful.[21] Such optimism remained a leitmotif for the remainder of the century, guided by confidence in the technical mastery over nature that promised to provide social, commercial, and material benefits to the empire and its inhabitants.

International negotiations and engineering works in the 1830s and 1840s also raised expectations that both Danube regulation projects and international trade would accelerate. Once Russia gained control of the Danube's sole navigable branch leading to the Black Sea in the Treaty of Adrianople in 1829, István Széchenyi pursued plans to regulate the lower stretches of the Danube, which he saw as a critical economic corridor.[22] Széchenyi's diplomatic outreach to Ottoman, Serbian, and Wallachian authorities in the early 1830s smoothed the way for Pál Vásárhelyi's plans to clear the rocky cataracts of the Iron Gates (1833–36).[23] Once the regulation work began, the administration of the Danube Steam Navigation Company (DDSG) thanked the imperial authorities for supporting the expansion of steam navigation into the Lower Danube and praised Franz I's efforts for having "commanded the greatest haste in securing a favorable outcome in this matter."[24] Further conversations with the eminent engineer József Beszédes also convinced Széchenyi that regulating the Lower Danube would help drain floodwaters faster from the river's upper stretches in Hungary, thus reducing flood dangers.[25] While the rapids had only been partially cleared by 1836, the concomitant construction of a portage road next to the river did facilitate greater trade and mobility along the river.

These efforts were aided by improved international conditions. Free navigation agreements between the Habsburg Empire and the Ottoman and Russian Empires in 1838 and 1840, respectively, meant that ships could sail on foreign stretches of the Danube without paying burdensome dues. Upstream in Bavaria, construction workers started on Ludwig II's ambitious Ludwigkanal in 1836, which envisioned uniting the Danube with the Main (and thus Rhine) Rivers. Even before work began, many publications argued that such a project would transform the Danube into a major artery of freight and passenger traffic from the North Sea to the Black Sea.[26]

While river regulation works grew in scope and ambition, they suffered domestically from chronic underfunding. Salt tax revenue in Hungary remained the main revenue to fund public works like bridge construction, while the funding and management of river regulation works was left to local landowners considered the primary beneficiaries of such projects. In the 1830s and 1840s, reformers heavily criticized the salt tax revenue as woefully inadequate for the growing scale of public works. Both Lajos Kossuth and István Széchenyi targeted this funding method in their daily newspapers *Házadó* (Domestic Tax) and *Az adó és két garas* (Tax and Two Coppers), respectively. Both pleaded for capital-based funding and criticized the government's backward taxation system, which exempted aristocrats from paying. Private companies also complained of governmental neglect on the waterways. Representatives from several flood-prone counties in Hungary petitioned the National Diet (Országgyűlés) to regulate the Danube, Lajta/Leitha, Rába, Rábca, Körös, Berettyó, and Tisza to no avail. Robert Nemes details how the Hungarian engineer Pál Vásárhelyi was forced to prematurely end his work regulating the Iron Gates rapids after Emperor Franz's death in 1835 when authorities in Vienna found it difficult to find the funds for it (and some felt Vásárhelyi was clinging too firmly to national rather than imperial principles for the projects). Landowners too were fickle funders of regulation works. Having returned from the Lower Danube, Vásárhelyi next sought to rouse financial support for the Tisza's regulation from ambivalent landowners, who appreciated increasing their arable landholdings but were critical of the projected costs of the plans.[27]

This problem was hardly unique to Hungary. The Bohemian, Austrian, and Italian crownlands each had their own engineering directorate (*Baudirektion*) or administration (*Bau-Verwaltung*) responsible for civil engineering projects. Except in Lombardy, Venice, and Hungary, provinces' engineering departments taxed local populations directly to raise specific funds to run their administrations and complete projects.[28] Each department's budget was divided into various expenditures including administrative costs, maintenance costs for existent hydraulic works (repairing embankments and bridges, dredging waterways), and undertaking of new hydraulic projects (constructing embankments and bridges, blasting cataracts, digging transections). The imperial government recorded stagnating budgets for new engineering works on the Danube in the 1830s and 1840s. Meanwhile, a lack of any coordinating or overarching plans for regulating the Danube caused crownlands to struggle to make anything other than minor changes to the empire's rivers.[29] In 1842, the DDSG began recording the company's not insignificant expenses associated with its own, private efforts to keep the Danube cleared, even calculating the cost that unregulated rivers and resultant low water levels caused its ships' schedules and business.[30] A German work published in 1840, titled *Panorama of the Austrian Monarchy*, pointed out the continued hindrances to navigation on the Danube, arguing that shallow stretches froze sooner and unregulated stretches branched ceaselessly into side channels in lowlands and accumulated sandbanks and shipwrecks, which imperiled navigation even at average water levels.[31]

Furthermore, progress that was made led at times to unintended consequences. Regulation plans were designed to eliminate meanders and reduce a river's width, thereby increasing its speed and depth. An Englishman sailing down the Danube in 1839 described such works near Pozsony/Pressburg:

> Well-constructed embankments erected at a great expense a little below Presburg [*sic*] ... [meant that] the force of the current is turned in a

particular direction, and made to act on a fixed point with such power, that in a wonderfully short time it cuts out passages, brings down banks, straightens the course, and silts up whole arms, which would otherwise consume the water, and often lead to a change of the bed of the river itself ... by means of these embankments ... it is calculated that in a very few years, the course of the river will be straightened, its bed deepened, and the navigation rendered practicable at all seasons.[32]

This may have been the desired result, but it did not always happen. Severin Hohensinner has detailed how locals in the eastern Machland, a region downstream from Linz, Austria, built a training wall in the 1820s and 1830s to stop the Danube from meandering southward. Consequently, it deflected the river's current northward, threatening to inundate communities and agricultural land on the opposing bank. Crews had also excavated a canal through the river's winding course to make a permanent shipping lane in place of its shallow meanders. Water rushing through the new canal eroded its banks, sending sediment downstream where it settled in the river's slower stretches. This accumulated sediment forced the river to expand to nearly a kilometer in width. Both interventions, the training wall and the newly dug canal, thus made the river more unpredictable and failed to concentrate it in a narrow, reliable course. Work to fix these unintended consequences took decades, after which time the faster current transported more bedload, requiring constant dredging, while the narrower and deeper bed lowered the water table, drying out adjacent lands.[33]

In this climate of limited funds, regulation plans competed to remain germane. While a massive flood in 1838 led the Hungarian National Diet to pass legislation in 1840 to fund the Danube's regulation near Pest-Buda, by the mid-1840s, mapping of the Tisza and the more pressing work of its regulation siphoned away funds and attention. As Klára Dóka has discussed, even regulation of the Tisza, ostensibly the most important endeavor in Hungary, was caught up in the political schisms in the Reform Era between the more moderate, large landowners (upper nobility), and the more radical reformers (middle nobility).[34] With the support of Chancellor Metternich and Palatine Joseph, István Széchenyi had joined the conservative government of Count György Apponyi in 1844 as chair of the Transportation Commission. In this role, he served as commissioner for the Tisza's regulation, where he went about advancing a holistic plan to regulate the river and reclaim over 1,800 square miles of swampland (an area larger than the US state of Rhode Island but a bit smaller than the country Trinidad and Tobago).[35] In exchange for large landowners' support of his vision, Széchenyi propped up their interests, which came at the expense of the private, provincial water regulation companies championed by local middle-nobility reformers.[36] All this managed to slow efforts at implementing a comprehensive plan to regulate the Danube.

Change Inundates the Banks in 1847

From the last third of the eighteenth century to the mid-nineteenth century, winters in Central Europe were prone to shifts from colder to milder temperatures. In the 1830s and 1840s, they became decisively colder during the last iterations of the Little Ice Age.[37] With persistent colder conditions, ice frequently formed on the Danube. When warmer weather thawed the ice, it broke apart and floated as large rafts downstream. Where these so-called ice floes got stuck in shallow, unregulated stretches of the Danube, they formed ice dams that blocked the river and caused water to overtop embankments. In winter 1847, these conditions created massive flooding that overtook residents on the Danube.

The aftermath reawakened the dormant spirit of regulation. Once the waters subsided, Széchenyi wrote to Palatine Joseph arguing that public confidence would decline if the government did not speed up regulation.[38] The following year, the Engineering Directorate presented Széchenyi with regulation plans they had been drafting for two years. The National Diet expressed regret that it had hitherto considered the Danube's regulation a private affair rather than one of the public interests.[39] Members of the public certainly felt the government was shirking its responsibilities in this regard. The Viennese Association for Friends of Natural Sciences meeting in winter 1847/48 drew prescriptive conclusions about the work that they expected the imperial government to undertake on the Danube. The group homed in on ice formation on the Danube, which had caused the worst winter floods of the past two decades.[40]

The prescriptions of the Association for Friends of Natural Sciences exemplified one of the ways in which people had come to depend on and expect government support for their well-being. Since the mid-1750s enlightened reforms emanating from dynasty, its ministers, and the imperial bureaucracy had sculpted Habsburg society, forging a "eudaemonic state" to secure the population's general well-being.[41] Only the extremes of the French Revolution had tempered Franz II's enthusiasm for liberal concessions, though it had not stopped him from pursuing legal and civil reforms, notably the *Allgemeines bürgerliches Gesetzbuch*, which homogenized and rationalized the legal code among the empire's citizens. In the *Vormärz* period, however, Emperor Franz assumed the state could operate sufficiently without additional intervention.[42] Consequently, reforms stagnated.[43] As social and economic conditions worsened, others in society were less content with peaceful discussions about the government's role in society. The massive 1847 flood had exposed the Danube's incomplete regulation and catalyzed growing impatience with the imperial authorities. When uprisings broke out in Italy and France in early spring 1848, people in the Habsburg Empire followed suit.

There is no denying the national and liberal tones of the 1848 revolutions. Some protesters agitated for national autonomy while others demanded constitutional reform. Most expressed a desire for a reformed, rather than deposed, Habsburg state.[44] Thus, it is possible to understand the uprisings not as a *counter* to the imperial government's rule—except for revolts in Italy and Hungary—but as a desire for increased responsiveness to the public will. Given the lingering pangs from the 1847 flood, the government sought to use the Danube as one tool to address grievances and redress the ailing populace.

Once Franz Joseph I ascended to the throne and quelled the uprisings, he issued a series of reformative decrees to placate the population. A new, centralized bureaucracy in Vienna deployed massive public spending programs to improve economic, commercial, and financial conditions and dispel (or at least mitigate) discontent. Showcasing efforts in the official organ *Austria Daily Paper for Trade and Industry, Public Works, and Transportation*, the Imperial-Royal Trade Ministry devoted its opening edition in January 1851 to an explanation of these new reforms. Replete with platitudes and government slogans, it highlighted a recurrent theme—that the "vigorous regulation of rivers and canals" and promotion of steam navigation joined the multitude of political and economic reforms, engineering works, and technical innovations that contributed to greater unity.[45]

Words of "unity" may have fallen flat in the politically repressive era following the 1848 uprisings, especially among the Hungarians, but the 1847 and 1849 floods undoubtedly demonstrated to the empire's residents that the unregulated Danube would only continue to inflict damages if not addressed. It is telling that in the midst of 1848 revolution, the new Hungarian cabinet, for example, had felt compelled to assure the masses that it intended to expand public works on the Danube.[46] It was clear to those suffering repeated floods that a unified vision for the Danube's regulation was indeed needed. Franz Joseph's neoabsolutist reforms founded a single engineering body in Vienna, the General Baudirection, which took over managing all provincial engineering departments.[47] For the first time, a single organ coordinated all hydraulic engineering projects in the empire, including direct control of Hungary's hydrological affairs. In Hungary, it directed affairs from the Buda-based National Engineering Directorate (Országos Építészi Igazgatóság) run by Viennese-appointed directors. This office coordinated the technical administrations in each of the five military districts (Pest, Sopron, Pozsony, Kassa, and Nagyvárad) set up by the army after the uprisings. Each of these five administrations had its own engineers, assistant engineers, and trainees charged with overseeing regulation works. While the five district engineering offices were similar in composition, they differed in the scope of their work; some focused more on civic construction and roadbuilding and others on river regulation.[48] Bolstering the operations of the district-level technical bodies, the imperial authorities also set up River Supervisory Offices (*folyamfelügyelők*) manned by local river engineers along the empire's most prominent rivers to monitor levels for shipping purposes.

This consolidation of political-technical authority provided a key ingredient in the government's plans to shepherd a more comprehensive plan for the Danube's regulation. This had strong parallels to the Rhine's regulation earlier that century, when Napoleon's forcible mediatization of hundreds of German states vastly reduced the number of riparian states along the Rhine. With fewer states involved in negotiations, the engineer Johann Gottfried Tulla was able to commence his far-reaching plans for the Rhine's regulation, described by David Blackbourn as "the largest civil engineering project that had ever been undertaken in Germany."[49]

The centralized administration in Vienna enjoyed a more streamlined approach to regulation. With a greatly enhanced budget, it dispatched engineers to once again focus on regulating and improving navigation on the Danube.[50] Franz Joseph personally commanded the blasting of the *Wirbel* and *Strudel* in the 1850s, ostensibly after they

The Danube as Life Artery

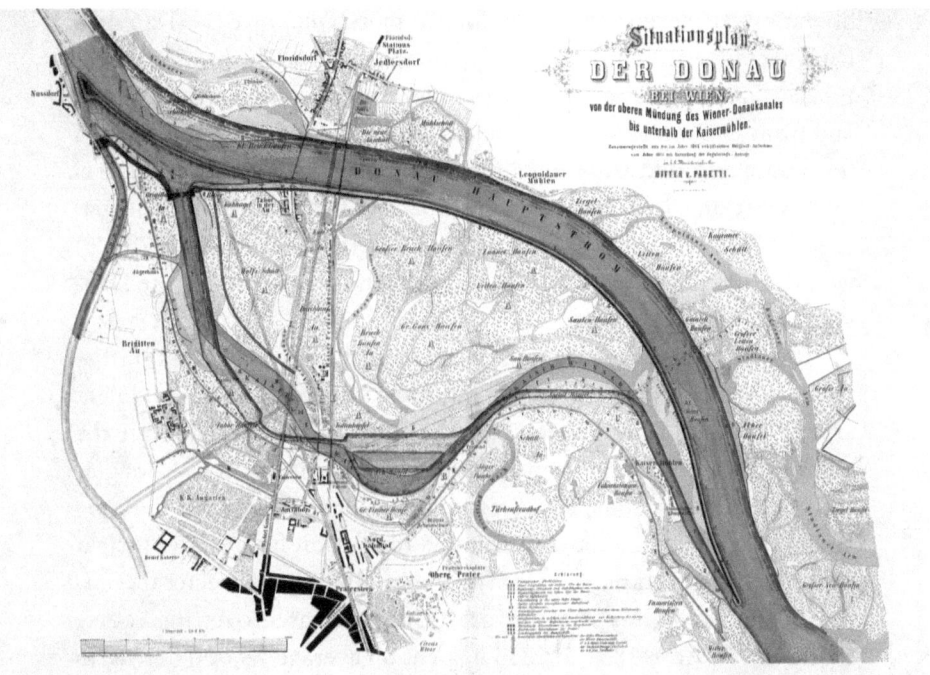

Figure 2.1. Florian Pasetti's Regulation Plans, 1862 © Wien Museum.

had endangered the ship of his soon-to-be bride Elisabeth during her 1854 trip down the Danube for their wedding. A decade later, the notable scholar and statistician Carl Freiherr von Czoernig, who had spent the 1850s overseeing efforts to improve the Danube's navigability, expressed confidence that these two obstacles would soon be dangers "only living in memory."[51] Work in the 1850s and 1860s also focused heavily on containing the meandering and branching river within a defined bed using leading walls, transections, and groins, and securing banks with trees and stone embankments. Of the Austrian stretches of the river, nearly 50 percent was finished by 1865.[52]

Where opinions diverged, progress was more ambiguous. Following the 1847 and 1849 floods in Vienna, a new regulation commission in 1850 disagreed on how best to regulate the river near the capital. The river's serpentine and braided nature made efficient regulation difficult, as each of the braided arms was governed by its own hydrological conditions, effectively requiring engineers to regulate multiple rivers at once.[53] Unfortunately, unlike the consensus for removing hindrances like rocks and sandbanks, the regulation committee's deliberations disagreed about whether to merely *improve* the Danube's natural course at Vienna or dig a singular, new bed. The head engineer Florian Pasetti's design to retain a more natural riverbed prevailed, and from 1850 to 1861, the Danube's regulation involved closing off a few branching side arms and strengthening a single, existing channel for navigation (fig. 2.1). A flood in 1862 revealed the plan's inadequacies. Only later, during the great regulation (undertaken in 1870–75) would workers and engineers excavate a whole new bed for the Danube at the imperial capital.

Regulation work in Hungary proceeded with collaboration between engineering directorates and local companies and landowners.[54] Negotiations stipulated that the state budget would cover the costs of regulation and keep rivers cleared for navigation as long as local landowners and private companies shared costs for building embankments. While precise Danube regulation figures are unclear from the official statistical reports, the Engineering Directorate's expenditures on hydraulic works drastically increased over the course of the 1850s: from 1852 to 1855 it spent on average 28,000 florins on new works, which increased to over 500,000 florins on average in the years 1856 to 1859.[55] The semiofficial *Budapesti Hírlap* provided an overview of progress in the mid-1850s. Iterating the often-repeated refrain that the Danube and its regulation was a crucial part of imperial unity, one article complimented the Imperial-Royal Trade Ministry for undertaking such "patriotic" work and employing a more unitary vision, which earlier regulation endeavors had lacked.[56] The *Hírlap*'s pro-regime sentiment may not have resonated strongly with Hungarian readers, but the prominent liberal-nationalist politician Ferenc Deák, a hero to most Hungarians, set an example for many by tacitly accepting the government's policies to develop the imperial economy and eschewing appeals for political disruption.

At the end of the nineteenth century, the prominent Hungarian engineer Béla Gonda took stock of the Middle Danube's regulation work in the 1850s. From his vantage point, he argued that the regulation work had few overarching plans and instead dealt ad hoc with the "needs of the moment." He claimed that despite regulation projects, such as closing off side arms and dredging sandbanks, there were few permanent changes to the Danube until after 1885.[57] Dóka has criticized the bureaucratic reorganization during the neoabsolutist period, arguing that the "frequent exchange of personnel and reordering of offices top to bottom did not favor construction work."[58] In another work, she pointed out that many companies working on the Danube were formed by local interests and communities, and they sometimes worked at cross purposes.[59] Even Pasetti's report from 1861 acknowledged that the Danube's regulation between Gutor and Szap, which the Hungarian National Diet had approved funds for in 1831 and which Széchenyi complained were still incomplete in 1847, were likewise still under construction even in the early 1860s.[60]

Officially, the regime and its representatives continued to tout their commitment to regulating rivers and expanding waterway traffic throughout the empire. In 1850, the *Wiener Zeitung* published remarks by the trade minister announcing the imperial government's significant investment in the Tisza's regulation, managed by a stock company, to reduce flooding, reclaim land, and above all improve navigability.[61] Similar investments were made throughout the empire, as hydraulic engineering directorates in each province reported regulation and dredging projects on the Danube and its tributaries, such as the Tisza, Traun, Salzach, Enns, Inn, Mur, Drava, Boldrog, Körös, and Sava. Czoernig, recognizing the slow start to regulation in the 1850s and the disruption during the Crimean War (1854–56), nevertheless claimed in 1858 that "despite recent circumstances, which have hindered the state's ability to reach its full strength, there was no period in which so much has been done to improve water transportation than in the last seven years."[62] War with Piedmont in 1859 once again bankrupted the imperial coffers and led to a temporary rollback of public works funding. Franz Joseph's constitutional reforms won back the wealthy bourgeoisie, and hydraulic engineering directorates' budgets increased from 1860 to 1865.[63]

Some were concerned with the ecological consequences of large-scale hydraulic projects. Deborah Coen discusses the life and work of Anton Kerner von Marilaun, a naturalist whose observations of swamps and wetlands in Hungary led him to believe that they served a moderating influence on the temperature swings (and thus climate) of the Hungarian plains, the same way that people believed forests did elsewhere. He thus feared that large-scale drainage projects, such as those in the Tisza Valley, would transform the former floodplains to marginal, arid land and subject peasants to inhospitable growing conditions. As Coen indicates, the solution was obvious to Kerner: the imperial authorities had the duty to excavate irrigation canals across the plains and construct reservoirs to trap floodwaters to prevent the permanent disappearance of water in the region.⁶⁴ Zsolt Pinke has argued that Kerner's fears came to pass within a few decades. Landowners controlled the Tisza's regulation and drainage in the Tisza Valley and profited from the results: more land, more grains. When climate shifted and brought drier years after the 1840s until the mid-1870s, small landowners and those engaged in animal husbandry were vulnerable to extreme droughts and famine.⁶⁵

Unofficial evidence indicates that regulation itself was a slow process that did not dramatically alter certain stretches of the river for decades. When the German ornithologist Anton Fritsch wrote about his birdwatching and hunting trip on the Danube in 1852, he listed several dozen bird species that he encountered in the "swampy" lands adjacent to the river downstream from Pest-Buda. When attempting to observe certain roosting spots on a few islands in the Danube, he noted that one was "so overcrowded with willows and had such waterlogged banks" that his ship could not approach it.⁶⁶ His descriptions painted the Danube less as a highly regulated riverscape and more of a semi-wild space where birds appeared unthreatened by any enormous disruptions to their watery homes. According to another ornithologist, the Bavarian pastor Andreas Johann Jäckel, migratory birds, such as the thick-kneed bustard, also benefited from unregulated segments of the Upper Danube, settling for the summer on islands that "frequently formed" from gravel settling on the riverbed wherever the Danube's course was "irregular."⁶⁷ Olaf Bastian and Arnd Bernhardt agree that, broadly speaking, mid-century Hungary was still characterized by a large variety of biologically diverse habitats amid standing water and multiuse water systems, which remained undisrupted until large-scale water engineering works in the 1880s.⁶⁸ Even when Klun spoke before the Imperial-Royal Geographical Society to show off the newly mapped Danube and Tisza Rivers, his words pointed as much to the positive results of regulation works as to the lofty aspirations of future progress. In 1879, the Habsburg crown prince Rudolf went on a birding excursion along the Danube, and his account suggests evidence that regulation was slowly changing the waterscape. In some descriptions, published later in the *Journal für Ornithologie*, some segments of the river remained characterized by branching arms and swampy and willow-covered banks. But at Adony Island, which had previously been covered in poplars, swamps, and an impenetrable underbrush, workers had cleared it of trees and drained its banks, so that "it was neither a primitive jungle nor wilderness and no longer an *Eldorado* for birds like in earlier times."[69]

The slow pace of change did not dampen the passionate support for regulation held in some quarters. Many remained convinced, or at least vocalized the belief, that the entire empire benefited from these projects. The influential *Pesti Napló* passionately argued in September 1866:

> Danube navigation is so important not only for our country, but for the trade of the entire empire ... the need for the Danube's regulation is so widely acknowledged as a requirement for the promotion of domestic and imperial prosperity that it would be unnecessary to discuss. At the imperial-royal state engineering office, I believe there are perfect designs for the Danube's regulation, the only motivating force that has been lacking so far has been the money. It is true that regulation will be giant work and cost millions, and yet there is hardly a better time than now to undertake it.[70]

This image of the Danube and its regulation underpinning imperial unity came at an extraordinary moment in the empire. The imperial army had just been humiliated by the Prussians in the Seven Weeks' War in June, and a few months later, in February 1867, Franz Joseph finalized and approved the Austro-Hungarian Ausgleich. Dividing the empire's internal governance between Vienna and Budapest in two halves referred to colloquially as "Austria" and "Hungary," the Ausgleich or "Compromise" doubled the government institutions working on Danube affairs. In Austria, the Trade and Interior Ministries took responsibility for Danube and general river regulation. Authorities in Budapest established a new Ministry of Public Works and Transport (Közmunka- és Közlekedési Miniszterium), which oversaw construction projects like embankments, steamboat stations, and bridges. These ministries retained the power to suggest legislation and request budgets for their particular projects and were responsible for overseeing hydraulic engineering work.

According to the legal terms of the Ausgleich, the ministries and parliaments in each half of the empire were only obligated to cooperate in the affairs of the three joint ministries: finances, foreign affairs, and the military. Nonetheless, shared concerns, such as establishing a common Danube policy, drove dialogue and cooperation between both halves of the empire as well. The ministries charged with river regulation found ways to work together after the Compromise. By December 1867, the legislative bodies on both halves of the monarchy signed an agreement for a continued customs union, which Franz Joseph declared as law on December 24. The new law provided a joint framework for trade, which then prompted both ministries to consider common navigational guidelines as well. Representatives of the Imperial-Royal Trade Ministry wrote to bureaucrats in the Ministry of Public Works and Transport and discussed the need to craft a law that would govern inland shipping and navigation within the monarchy. To harmonize their approaches, each ministry requested examples of the laws the other ministry was crafting. Securing these drafts involved coordinating several government agencies and committees, sometimes taking months and leading to exasperated memos and requests sent back and forth. Despite hindrances, intraministerial communication functioned well enough to collaborate on and address common issues that promised greater economic development or trade within the entire empire. A more significant change came from the new avenues of communication and scrutiny from the public that accompanied the new constitutional arrangements after 1867, as legislative bodies, petitions, and burgeoning civic engagement all demanded action on regulation.

Actors and Advocates

Engineering the Danube had always required the input of technical experts, and plans had frequently weighed the relative merits of technical as well as commercial or political benefits.[71] As far as the public or private industry was concerned, the government tried to promote policies that favored commercial growth and cohesion (see chapter 3 in this book), take into account how certain hydraulic projects would protect people from flooding (see chapter 4), and mitigate tensions between different stakeholders along the river (see chapter 5). In the era after 1867, new parliamentary systems of governance and ministries devoted to public works projects provided even greater opportunities for civic engagement as members of the public wrote to the government and mobilized other members of society to make clear how river regulation affected them and what it meant for the empire.

Navigation companies and river engineers strongly favored greater government intervention to both fund and oversee projects. The Danube Steam Navigation Company requested help clearing hindrances on the Tisza and its tributaries as it expanded steam navigation upstream.[72] Requesting help for the Maros River's regulation, maintenance, and improved navigability, an engineer in Arad claimed that this investment would benefit the surrounding countryside and be in the interests "of the whole country."[73] Local county officials frequently requested help in removing hindrances to certain rivers, pointing out that the unregulated stretches froze too readily, hindering its navigability.[74] The government was a necessary component for accomplishing these tasks. After having studied boat traffic on the Danube from 1849 to 1869, Johann Winckler concluded that state investment would ensure that the river remained "the most important life artery for transporting particularly mass goods … the state just had to keep it in better condition as a shipping lane and build landing places and transshipment sites to grow riverine trade."[75] This sometimes drove central authorities to intervene at the expense of local companies. In 1873, a company formed to regulate the Rába River, a tributary of the Danube. However, engineering plans were so expensive and complex that the company was unable to start work. After much delay, the Hungarian government passed a law in 1885 forcing the project's local financiers to take out a loan to pay for the expected costs and empowering a royal commissioner to oversee the work, which the company then completed by 1893.[76]

Nongovernmental organizations played a large role in improving public knowledge and mobilizing public opinion about the Danube's regulation, the issue of improving waterways, and the construction of artificial canals. In Hungary, the Pest Chamber of Commerce wrote to municipal authorities and the National Diet to emphasize that neglecting river regulation affected trade.[77] In Austria, groups like the Lower Austrian Business Association, the Viennese Trade and Business Association, and the Upper and Lower Austrian Chambers of Commerce had tens of thousands of members, to which they wrote to inform them about economic issues linked to regulation works. On occasion, they prompted their members to vote and take more proactive measures to promote river trade. Chambers of Commerce were one of only four curia or estates that elected deputies to the House of Deputies (along with rural communes, large landowners, and cities) and their advocacy and occasional petitions to the whole house wielded practical influence.

Individual experts also took a leading role in public education and engagement. As the Danube Association secretary, member of the Viennese communal council, deputy to the Reichsrat, and professor, Eduard Suess spread the Danube gospel in as many commercial, technical, and governmental proceedings as possible. In his speeches and publications, he argued passionately that overcoming nature's obstacles to progress, particularly the Danube's regulation, would enrich every city along it and create a "peaceful Danube policy" that would succeed in "spread[ing] culture and multiply[ing] the domestic and foreign well-being."[78] He made clear to his audiences that the Danube's regulation was a joint responsibility in the empire. When a colleague in the House of Deputies alleged that the Hungarian government was shirking its responsibility by not regulating the Iron Gates faster, Suess defended the Hungarians' efforts and explained from his own firsthand experience at the site how difficult the undertaking was.[79]

Other technical experts frequently waded into the issue, expressing a general confidence that hydraulic engineering would ultimately strengthen defenses against floods and broaden the river's utility, be it for navigation, irrigation, or industry. Viktor von Domaszewski, a hydraulic engineer, published several works at the end of the 1870s touting the necessity to regulate rivers, which otherwise ran amok, or to drain alluvial floodplains, which threatened settlements.[80] Hungary's prominent engineer Károly Hieronymi, who served as a secretary in Imre Mikó's Ministry of Public Works and Transport in 1868, gradually published works ranging from a general view about public projects and transportation to a specific focus on river regulations by the 1880s.[81] Louis Zels, the eventual editor of the Danube Association's weekly magazine *Danubius* from 1886 onward, published several pieces dissecting the costs and benefits of regulation with regard to navigation, as well as the general competitiveness of waterways vis-à-vis rails.[82]

These opinions fell on fertile ground, and strong public support galvanized government funding in the 1880s and 1890s along the Danube and its tributaries. In June 1882, both the Reichsrat and the Lower Austrian Landtag approved the Danube's entire regulation in Lower Austria.[83] In 1885, the National Diet in Budapest passed Law VIII providing 17 million florins to regulate the Danube in Hungary.[84] In relation to prior budgets, 17 million florins compared very favorably; the Engineering Directorate in Buda had previously spent 4.5 million florins in total from 1850 to 1888 on the Danube's regulation.[85]

Danube tributaries likewise received imperial and royal funds. The Upper Austrian authorities also approved regulation of their provinces' tributaries in the 1880s and 1890s, projects they viewed as warranting imperial support. The Upper Austrian Landtag voted to petition the imperial government for funds to regulate various rivers, including the Traun River (1884), Ager River (1893 to 1896), the Asch River at Mining and Mühlheim, the Traun at Ebensee, and the Inn River (1892).[86] Gábor Baross's "iron ministry" endeavored to improve transportation and commerce in Hungary, thus prioritizing infrastructure spending. In early May 1893, the House of Representatives Water Affairs Committee submitted a legislative proposal "regarding the Middle Danube's complete regulation as well as the necessary regulatory work on the country's other significant rivers."[87] By December, both chambers approved 20 million florins to regulate the Danube and 31 million florins to regulate a dozen other rivers in Hungary "urgently necessary from the perspective of commerce, navigation, and flooding."[88] King Franz Joseph provided the final ratification.

The Danube's regulation thus had far-ranging consequences within the empire because it set the expectation that other rivers would also be regulated, regardless of their length or size. Given the empire's river-rich nature, this meant that populations living far from the Danube could nevertheless expect imperial attention and support for local rivers, streams, and waterways. These projects were possible because an engaged public and enterprising group of technical experts and commercial interests worked to move the levers of government. As funding flowed from central authorities to local bodies to help regulate rivers, they dovetailed with other plans taking aim at a more challenging feat for the mountainous empire: building out the network of artificial canals to unite these rivers into a greater, cohesive whole.

The Canal Issue

In 1869, Franz Joseph traveled to Egypt to attend the opening of the Suez Canal. French capital and Egyptian laborers (and later steam excavators) had driven the massive engineering feat, one of the largest at the time. While initially resisted by the British, the 164-kilometer waterway soon transformed Britain's cultural ideas and logistical plans for their empire, with upswings in steamship construction, trade, and communication tightening its hold over colonies in South and East Asia.[89] The Habsburg Empire also had a historic tie to the project; when the Egyptian ruler Mohammed Ali was contemplating the project, he had approached Austrian Chancellor Klemens von Metternich for advice. Even after he was ousted from the imperial government, Metternich remained closely involved with plans for the canal.[90] Franz Joseph's favorable impressions and reports of the Suez Canal back home fueled discussions in the Habsburg Empire regarding its own capabilities and plans regarding the "the canal question" (Ger. *Wasserstrassefrage*; Hung. *csatornakérdés*).

The idea of excavating artificial waterways around the empire was hardly new. Imperial authorities had built several large canals, the Franz Canal, Wiener Neustädter Canal, and the Franz-Joseph Canal being the largest and most notable. There were also many unfulfilled plans from Maria Theresa's vision to build canals allowing direct transportation from Vienna to the Adriatic Sea or schemes to connect the Moldau and Danube via a lock-and-dam canal system.[91] However, the Suez Canal's opening prompted a more serious consideration of the financial and technical requirements to pursue these plans. In 1869, a white paper enumerated the many compelling reasons to build a Danube-Sava Canal, including looking at the historical trends in traffic on the river and the estimated cost savings it would provide. The same year a technical report was published detailing the geographical and hydrological conditions of the proposed route and estimating the technical interventions that would be necessary to build the canal.[92] In May 1872, the deputies in the Reichsrat debated legislation that would allocate funds to construct a canal between the Danube and Oder Rivers. One newspaper, *Vasúti és Közlekedési Közlöny* (the Railway and Transportation Gazette), projected that the canal would increase domestic river traffic and should therefore be "happily welcomed."[93] While most deputies seemed inclined to support the law, lingering concerns about financing the canal construction turned out to be the most prescient. Two months after the law's proposal, overspeculation caused the Viennese stock market to collapse, and with it the plans for a Danube-Oder Canal.[94] The canal idea did not disappear. The following year, another representative submitted a petition to the Reichsrat, and interest also remained in Hungary.[95] While the political realm held canal legislation in abeyance, publications from 1873 onward kept the spirit alive and laid the foundation for renewed canal legislation at the turn of the century.[96]

In this regard, interest in canals and water transportation reflected the European zeitgeist in the last decades of the nineteenth century. The mid-century transition to rail travel had overshadowed traffic on the smaller canals built in earlier centuries. But by the 1870s and 1880s a renewed interest in canal construction, driven by favorable projections for shipping costs on waterways compared to those on railways, gripped Continental Europe. In 1875–85, the German Empire's waterway traffic increased by 66 percent, in 1878–88 France's increased by 50 percent, in 1876–88 Russian grain exports on rivers increased from 380,000 tons to 790,000 tons, the Rhine traffic alone in 1890–91 increased 80 percent, the Main traffic at Frankfurt in 1882–90 increased from just over 9,000 tons to 1.2 million tons, and traffic on the Berlin Spree carried 50 percent of citywide traffic.[97] Legislation in France, Russia, and Germany from the 1870s onward focused on developing large-scale canal networks to connect existing rivers. In 1883, Germany had 7,000 thousand miles of navigable waterways. After Germany's unification, new projects added another 1,000 miles of new waterways by 1913.[98] In 1879, France had over 7,000 miles of waterways, including 3,000 thousand miles of canals, 80 percent of which were under state control.[99] After France's loss to the Prussians in 1871, the French National Assembly decided to update its older canals and vowed to spend 1 billion francs constructing an additional 1,800 miles of waterways, dwarfing the 10 million francs it had spent between 1814 and 1870.[100] In the late nineteenth century, the Russian Empire had 35,000 miles of navigable waterways.[101] Its 550 miles of canals permitted navigation between the Volga, Dnepr, Don, Dvina, and Ob Rivers.[102]

To coordinate Europe's waterway interests during the era of colossal hydraulic engineering projects, Brussels played host to the first ever Inland Waterway Congress in 1885, which brought together government representatives, commercial parties, and technical experts to meet and discuss how to promote river regulation, canal construction, and commerce on European rivers. Interregional cooperation was already taking place in the Netherlands, with some Dutch canals integrated into the river networks of several states. Germany, Switzerland, and Austro-Hungarian representatives organized their own annual congress in the late 1890s to coordinate technical, navigational, commercial, agricultural, ministerial, and municipal interests in Central Europe.

To educate audiences in the Habsburg Empire, professionals and state functionaries gave speeches to commercial, technical, and engineering groups enumerating the many advantages for the empire if it built a wider network of canals. In a speech to rail functionaries, Arthur Oelwein argued that waterways would always outperform rails in the transportation of mass and heavy goods and could more easily unite trade across other waterways across Europe, unlike differing rail gauges.[103] He frequently compared Austria's water-rich but canal-poor empire with neighboring states, once applauding France for investing in its canals after the humiliating loss to the Prussians, in order to strengthen trade, industry, and national development. In his numerous speeches supporting the expansion of waterways in the empire, Oelwein frequently found ways to declare that the Danube-Oder Canal's construction should be the empire's "highest order of business."[104] When some criticized his plans as too expensive, he ridiculed the notion that Austria was too poor to invest in waterways. Even assuming the government did have to take on massive debts to pay for canal construction, he argued that it would be a worthwhile investment.[105]

Plans eventually reached the upper echelons of government bodies. Following the 1867 Compromise, the governments in Vienna and Budapest met every ten years to discuss issues of common interest, such as conditions of the customs union and contributions to the joint budget for common finances. The 1896 joint negotiations brought up their common interest in expanding waterways, specifically, the Danube-Oder and Danube-Moldau-Elbe Canals.[106] The following year, the disastrous Badeni language crisis led to political gridlock in the Imperial Diet, but by early 1900, the imperial government once again took up canal planning with public enthusiasm behind it.[107] Dozens of books about canal construction, canal plans, the benefits of canals, and other more technical discussions flooded the markets.

To implement these plans, Emperor Franz Joseph asked Ernest von Koerber, a dedicated reformer, to form a cabinet and serve as prime minister in January 1900. Koerber was first and foremost a bureaucrat, who nevertheless was quite successful in passing legislature and making deals in office. In his four years in office, Koerber vigorously pursued reform of the imperial bureaucracy and the modernization of imperial infrastructure, including the expansion of rail and canal construction.[108] Robin Okey has pointed to Koerber's legislative record as evidence that technocratic rather than political leaders were more successful at finding common ground and working with diverse members of the empire at this time.[109]

The Danube as Life Artery

To build public enthusiasm for canal legislation to pass under Koerber's leadership, the Danube Association organized a "Wasserstraßentag" (Canal Congress) in December 1900. Regional newspapers advertised the "Wasserstraßentag," as did socialist, commercial, and engineering journals. The *Neue Freie Presse* surmised that the assembly would pressure the Reichsrat to invest 400 million crowns to channelize the empire's rivers and build more canals.[110] The imperial government sent delegates to the Imperial War Ministry and its Marine Section, as well as the Foreign, Finance, Railway, Agriculture, and Trade Ministries. Provincial executive and representative bodies, municipalities, and corporate bodies from Bohemia, Silesia, Lower Austria, Galicia, and Upper Austria all attended. Alajos Hoszpocsky from the Commerce Ministry and György Rupcsics, the Hungarian Navigation Association's vice president, represented Hungary's interests. Industrial, commercial, technical, and agricultural representatives reportedly found the canal proposals a good idea, and despite their political disagreements, the Germans and Czechs appeared unanimous in their support.

The assembly's success was immediate. On March 1, 1901, the Reichsrat's House of Deputies formed a Canal Committee to discuss the proposed canal plans and it made its way through the legislative process. Papers in Budapest took note of this process, particularly after Prime Minister Koerber announced the forthcoming canal legislation in March, causing the Viennese and subsequently the Budapest stock market to rise.[111] The Budapest papers urged Hungarians to take up this issue to ensure that they would be able to take advantage of the vastly augmented river traffic foreseen by Austria's new canals.[112]

In April, Koerber introduced legislation requesting funding for canal construction and river regulation. The proposed law envisioned canals uniting the Danube with the Oder and Weichsel, connecting the Elbe and Moldau, and regulating numerous rivers in Bohemia, Moravia, Silesia, Galicia, and Lower and Upper Austria. A line of credit of 250 million crowns (125 million florins) would be approved to be used exclusively for building canals, and the work had to begin by 1904 and be completed within twenty years.[113] Such a project planned to integrate Bohemian and Galician crownlands' major rivers with the Danube and the imperial core. In June, deliberation of the law began in the House of Lords and House of Representatives. On June 11, 1901, positive passage of Imperial Law 66 committed Austria to an ambitious improvement and integration of its waterways.

In the coming months, the new canal law elicited a wide variety of reactions. Initial reports in both halves of the empire were positive while later articles detailed more criticism of the law's ambitious and perhaps unrealistic scope.[114] Nevertheless, experts at the German-Austrian-Hungarian Inland Navigation Congress in Breslau expressed confidence in the technical ability of Austria's engineers.[115] When the imperial government had made little headway in finalizing plans by the following April, though, some Viennese papers voiced skepticism that the project would be implemented.

Part of the delay came from the empire's federalized structure and competing canal route preferences in the provinces. Because debates cut along geographical lines, cross-provincial interests united regions like Bohemia and Upper Austria, and delays came not from nationalist obstructionism but from the normal results of the democratic decision-making process. A provision in the imperial canal legislation set up a so-called Canal Council, which empowered provinces with both an advisory role and legal authority to enact the legislation. The Council consisted of half imperial and half provincial representatives from trade, industry, agriculture, forestry, and labor organizations, which together coordinated canal construction and river regulation. Although such measures were taken to include as many stakeholders as possible with the intent to identify and designate common ground, critics were occasionally vocal about the way decision-making and "general well-being" neglected some voices in the process. This was certainly the case a few decades prior with the *Wald-Klima-Frage* (forest-climate question)—the attempt of government officials and technical experts to address fears of climate change and forest coverage by outlining measures to protect and expand forests in the empire. At the time, critics argued that these plans benefited large landowners and railway companies and acknowledged that while forests purportedly addressed environmental issues like desiccation and flooding, they did little to amend inequitable access to forest resources.[116]

Canal construction planning likewise encountered disagreements, and the price of the democratic process was, as always, messy. Disagreements about the route were endless. The Upper Austrian Trade and Business Association, with counterparts from Salzburg and Innsbruck and the Action Committee for Canal Construction Linz-Budweis all petitioned for the Danube-Moldau Canal to enter the Danube at Linz rather than Vienna's suburb at Korneuburg. The Action Committee calculated that the alternate route would be one-quarter the price and would more promote the economic interests of more provinces, including Bohemia, Upper Austria, and Styria.[117] The Upper Austrian governor, Alfred Ebenhoch, urged members of the imperial government to consider these plans. Another representative from Upper Austria's provincial diet (Landtag) argued that taxpayers in Bohemia and the Alpine lands would only support a canal through Upper Austria.[118] Thus, canal plans in the Cisleithanian lands reflected the sometimes rancorous democratic process that the Habsburg Empire was hardly alone in experiencing.

At the same time that the Imperial Diet was finalizing canal legislation, Franz Joseph opened the National Diet in Budapest and took the opportunity to suggest that its members improve and expand the kingdom's natural and artificial waterways to solidify its economic and commercial footing.[119] One representative in the Képviselőház (House of Representatives) had brought up the "canal question" in March 1900, but despite his impassioned appeal to construct more canals, the body remained indecisive.[120] Discussions about canal construction in the new session followed many of the concerns that provinces in Cisleithania voiced, including that financiers of new waterways would not be primary beneficiaries.[121] Later debates acknowledged that the press "feverishly covered" the topic and the public was beginning to welcome it. Representatives and members of the government both emphasized that waterways would improve trade to the west and open up access to raw materials from the east, while the construction of adjacent drainage and irrigation canals would transform the sandy, salty marshes between the Danube and Tisza, open up fertile farmland, and improve property value.[122] These types of discussions continued annually over the next decade, though the National Diet faced a perpetual impasse regarding canal expansion, despite the general support for at least certain projects like the Danube-Tisza Canal. While successive government ministers remained strong advocates for advancing canal construction, representatives remained wedded to a policy of cautious indecision, ambivalently approving consistent funds to regulate rivers but delegating nothing to expand their reach. Despite negligent progress and waxing pessimism in certain circles, many still supported hydraulic projects, notably the king. Franz Joseph's speech opening the National Diet in 1910 again encouraged representatives to promote the general well-being of all, by passing legislation crucial to political, legal, and physical state-building, such as opening the franchise, respecting national minorities' rights, enhancing the public administration, and developing Hungary's economy by improving transportation on roads, rails, and waterways.[123]

In the last few years before World War I broke out, the imperial government tried to counterbalance regional peculiarities surrounding the "canal question" with designs for greater imperial integration. In 1910, the Polish Club introduced legislation in the Imperial Diet for a "Weichsel-Dneistr-San-Canal," to link Galicia's three largest rivers, enabling the province to develop an iron industry thanks to the flow of cheap coal. The imperial government's position was to subsidize the project and obtain ownership of the canal network via escheat after forty years. Galician provincial bodies and private funds would make up the bulk of the funding. These negotiations caused the pro-empire mouthpiece, the *Reichspost*, to claim that such inter-Galician canals would not benefit the imperial economy in the same way that a Danube-Oder-Dneister Canal would, something it claimed "even honest Poles would acknowledge."[124] The government tried to assuage both sides, publishing its analysis of costs, shipping data, canal profitability, and potential loss in railway revenue due to canal competition.[125] When rumors spread two years later in the *Lemberger Blatt*, a newspaper based in Galicia's provincial capital, that Kraków was suspending its canal construction, provincial authorities in Lemberg tried to reassure the local population that Galicia's interests in the Reichsrat still unanimously supported the projects.[126]

In Hungary, similar policy discussions were taking place. As in Galicia, particularism manifested itself in the inveterate concern that Hungarian waterways were not sufficiently cultivated to promote the *national* well-being. Discussions in the House of Representatives revealed a genuine desire to expand waterways and improve infrastructure such as the construction of transshipment stations to mutually support river and rail traffic. Several governmental ministries likewise responded by requesting money to regulate the Danube's tributaries linking resource-rich regions to other regions in need.[127] Many of these policies went against the tide of opinion, which championed rails, not waterways, for the economy's vitalization. By 1912, the difficulty balancing regional interests led the imperial government to acknowledge the need to revise the 1901 Canal Law. One newspaper succinctly summarized the zeitgeist, saying that "although the public opinion and government were happily confronted with the canal idea, the politically viable moment had arrived too early. None of the public discussions included exact studies of the proposed projects, which would provide estimates for the profitability of the construction."[128] In other words, while the government and the public in most provinces supported the Danube's expansion as a life artery of the empire, a lack of planning rather than a lack of enthusiasm eventually stymied plans.

The Danube in Times of War

In January 1914, seven months before World War I began, the German Military Administration bolstered its military rationale for building up its network of canals and waterways by arguing that they would "secure food supply in the interests of the country's defense."[129] Once World War I broke out, the idea of waterways guaranteeing the state's autarky gained traction in the Habsburg Empire as well. Representatives in the Reichsrat speculated that the Danube's importance as a trade route would grow thanks to the military alliance with Bulgaria and the Ottoman Empire.[130] The British blockade of northern Europe also imperiled supplies to Germany and made the Danube a much safer supply route, one that Romania's entrance into the war on the side of the Entente powers threatened in fall 1916. With Romanian forces entrenched on the Danube's northern banks, a Hungarian newspaper claimed that people had taken the river for granted for too long, given its vital role shuttling around food and munitions that were "deciding the war."[131]

Despite the ongoing war, the Central Powers met several times between 1916 and 1917 to discuss the Danube's future role in their cooperation and reconstruction. A month after Romania had declared war on the Habsburg Empire, over six hundred delegates from Germany, Austria, and Hungary met in Budapest at the Danube Conference to discuss the future of Central Europe's economic and commercial integration. The day of the conference, the *Pester Lloyd* claimed that the forthcoming agenda "should lay a significant piece for the beginning of the peaceful, postwar work, the basis for which is the 'holy Danube' [*heilige Donau*]."[132]

At the conference, the delegates outlined a future industrial, commercial, and agricultural space—a Danube realm—developing after the war, for which all powers would commit to jointly investing in river regulation, transshipment facility construction, and canal expansion. The following summer, delegates met in Vienna to solidify plans. Newspapers around the empire covered the deliberations, and opinions about it filtered well outside the capitals. Esztergom's mayor Béla Antóny, for example, penned a piece expressing his thoughts on the conference's goals. Antóny argued that any future victory of the Central Powers must also underpin humans' "general well-being and prosperity," which in his mind included "Central European economic union, waterways, the right to vote, and the question of democratic state organization."[133] Intentional or not, Antóny's priorities matched those that Franz Joseph I expressed opening the National Diet in 1910: reconstruction and state-building required a firm commitment to democratic and legislative reforms that advanced commercial and geographical integration.

Even in war, members of the government and civil society in the Habsburg Empire formulated ambitious Danube endeavors, which they and their forebears had pursued for centuries. They relit the spark of cooperation and unity along the Danube *and* repositioned the river as the key to states' economic prosperity and material well-being. A presentation by Rudolf Halter, the imperial-royal chief engineering officer, in November 1917 praised the slight, but ongoing progress of hydraulic projects during the war itself; a river dam had been erected to provide water for the Danube-Oder Canal, the Galicians had broken ground on their canal connecting the Weichsel to the Danube-Oder Canal, and the Upper Elbe River regulation was underway.[134] What none of the delegates ever envisioned, however, was the possibility of losing the war or that the empire would collapse. Instead, they built a future for themselves along the Danube, hoping to turn the river into an artery assimilating people and regions across Central Europe.

Rivers as Modern Means of Integration and Environmental Well-Being?

The goal of Central European integration, even during the war, was a natural culmination in the centuries-long policies and agendas that strove to regulate the Danube, improve the navigability of its tributaries, and eventually strive to extend the reach of those waterways. As countless newspaper articles, commercial guides, associational presentations, parliamentary debates, technical reports, and other documents made clear, the Danube and its tributaries provided the empire with a natural network for connecting resources and people.

The Danube as Life Artery

Engineering projects that regulated the river also sought to support the public's well-being. Growing potential for trade and increasing protection against flooding seemed noble goals from the perspective of planners and government officials. Of course, ecological consequences were inevitable, as drainage and regulation disrupted traditional flora and fauna, and with them a whole host of livelihoods. These were sacrificed because the purpose of these engineering works—cohesion, prosperity, and security—seemed to better serve the "common good."

Today rivers and artificial waterways in Europe hold an oddly familiar place in the public discourse as they did in the Habsburg Empire. One EU working paper points out that "more than 30,000 kilometers [nearly 20,000 miles] of rivers and canals, connecting hundreds of important cities and industrial areas, traverse Europe."[135] These connections play an important and integrative role in European transit. They have also given a second life to ideas such as the Main-Rhine-Danube Canal and the Danube-Oder Canal; the former opened in 1992, and the latter remains a perennial talking point between the Polish and Czech governments. Proponents of these projects have wielded familiar economic arguments to justify their construction, particularly the possibility that these canals will attract global trade and investment for the countries they pass through.[136]

In a major divergence with the nineteenth century, the environmental and ecological benefits of relying on water transportation are also touted. The United Nations Economic Commission for Europe (UNECE) argues, for example, that "inland waterways provide a low-cost, low-carbon, safe transport option for freight movements in many key European corridors, taking lorries off the road network."[137] The European Environmental Agency agrees that waterways provide more fuel-efficient transport options than roadways, asserting that "freight transport by inland waterways and short-distance sea shipping should increase by 25% by 2030 and 50% by 2050," in order to achieve the European Union's decarbonization goals.[138] Advocates of the Danube-Oder Canal claim that the project will increase the adjacent regions' ecological resilience by making them less flood prone and more drought resistant.[139] Environmental activists are skeptical of these last claims, arguing that planners of the Danube-Oder Canal did not consult experts who would have cited the immense ecological costs of the proposed waterway, in damages to habitats and ecosystems and in increased flood risks.[140]

The Danube as Life Artery

A major challenge that exists today is climate change. Droughts caused by climate change are drying up Europe's waterways and imperiling plans that rely heavily on river-borne freight transportation to reduce carbon emissions.[141] Reduced water levels on Europe's major rivers also cost companies millions and even threaten tourism and recreation.[142] We see in these quandaries echoes of the frustration and hopeful optimism that companies, governments, and communities in the nineteenth century expressed as they set out plans for the Danube River. It will be crucial to see whether today's competing interests and agendas can come together to find common solutions that can protect those at risk and help those looking to increase prosperity and integration in Europe.

Chapter 3

THE DANUBE AS A PEOPLE NETWORK

In 1848, two different men wrote to the National Assembly gathered at Frankfurt am Main to express a few thoughts on the Habsburg Empire and the situation in Europe. One was historian František Palacký, who politely refused to attend the assembly, claiming that "as a Bohemian of Slavic descent" he had no business attending a body dedicated to discussing matters of the German nation. The other was Baron Ludwig/Lajos Forgách of the Imperial-Royal Ceremonial Bodyguards at Vienna, whose aristocratic family lived in Moravia but had its historical roots in Hungary. In their writings, both men acknowledged the National Assembly's desires to elevate the German nation, but they themselves offered a broader vision of security, prosperity, and unity for the peoples of Central Europe based around the Habsburg Empire and the Danube River. While Palacký argued that the empire needed to offer equal rights to all its citizens, he nevertheless emphasized that its existence was imperative for protecting smaller nations in Europe and conferring advantages through this cooperation. He stated:

The Danube as a People Network

You know that in the south-east of Europe, along the frontiers of the Russian empire, there live many nations widely differing in origin, in language, in history and morals—Slavs, Wallachians, Magyars, and Germans, not to speak of Turks and Albanians—none of whom is sufficiently powerful itself to bid successful defiance to the superior neighbor on the East for all time. They could only do so if a close and firm tie bound them all together as one. The vital artery of this necessary union of nations is the Danube. The focus of power of such a union must never be diverted far from this river, if the union is to be effective and remain so. Assuredly, if the Austrian State had not existed for ages, it would have been in the interests of Europe and indeed of humanity to endeavor to create it as soon as possible.[1]

For Palacký the Danube served to bring together and defend people, as it provided the basis for projecting the Habsburg Empire's internal unity and its strength abroad, particularly vis-à-vis the Russian Empire in the Lower Danube. Espousing a similar idea, Forgách dedicated his work *The Navigable Danube* to the National Assembly and explained in his introduction that the Danube's cohesive force as a commercial waterway "unified the people living within its basin." He argued that flowing through the middle of the "Austrian Monarchy" and "European continent," the Danube concentrated the powers of Central Europe and was the "strongest lever for the internal development of industry and one of the most powerful incitements for widening and elevating [the empire's] spiritual powers."[2] For both men, the power and cohesion engendered by the Danube was not simply an abstraction but centered on greater trade, transportation, communication, and connections. It is noteworthy that at a moment of "national awakening" in 1848, two prominent men of the Habsburg Empire would rebuff a strictly national vision for their state and instead hold up the Danube as a critical, transnational component for underpinning the well-being of the empire's people.

Chapters 1 and 2 have explored how Habsburg rulers, governments and representative bodies, and members of the public pursued hydraulic engineering works in their efforts to improve navigability, secure people's material well-being, and promote imperial unity. This chapter assesses how these works affected the interactions and interconnections of communities along the Danube. Scores of bureaucratic memos, commercial reports, petitions, and statistical studies detail how certain businesses, people, and livelihoods thrived while others declined in an era of improved navigation and greater traffic on the Danube. Throughout the nineteenth century, population growth, changing land use practices, market dynamics, trade policies, technical innovation, and shifting weather were all influential in shaping the network of people tied to the river. By the dawn of the twentieth century, successful middle-course regulation and the tapering off of the Little Ice Age's unpredictable climatic shifts had reduced the long-standing impact that weather had on navigational conditions. Instead, factors like urban population growth, imperial trade policies, and citizens' activism took a more prominent role in driving the manner and direction of river traffic. Antagonistic political rhetoric and economic tensions occasionally marred exchanges and interactions taking place across the empire. Yet they never supplanted the integrative role played by commercial and personal traffic on the Danube, which strengthened bonds between both halves of the monarchy up until World War I.

A Danube in Transition

In the early nineteenth century, traffic on the Danube looked much the same as it had for centuries with only fluctuations in quantity. Wooden rafts, barges, and boats carried salt, wood, fruit, grains, merchants' goods, and other wares downstream. Sailors feared rocky cataracts and treacherous whirlpools at certain known points along their routes. They also knew to stay alert for gravel and sandbanks, which formed and shifted each year faster than humans could dredge and more frequently than they could map.[3] Furthermore, winding channels hindered any predictable river depth for their boats. Regular passenger traffic arose at the end of the seventeenth century, with *Ordinarischiffe* taking people from Regensburg to Vienna once a week. In favorable conditions, the journey lasted one week. If the ships encountered low water, storms, fog, or wind, the journey might last up to twenty days.[4] Despite these obstacles, boats were, according to one author, the preferred method of travel, given that rivers were smoother, less prone to accidents, and more scenic than roads.[5]

This rationale made transporting heavier or bulkier items on the empire's waterways preferable as well, compared to the empire's underdeveloped overland routes. Whenever rain fell in Hungary, half the country became a morass. Only public roads leading to the coast or passing through the mountains could be used on a permanent basis.[6] It was, however, an arduous process transporting upstream. Teams of up to eighty horses towed boats upstream, spending on average five to six weeks to get from Pest to Vienna and, in favorable conditions, fourteen days to get from Vienna to Linz.[7] Laborers were forced to halt every time a team encountered a moored ship mill on the river that blocked their path. This method was reserved for the most valuable or necessary goods like salt, grains, and trade goods.

Encouraged by imperial and royal edicts, towns and villages along the Danube cut down trees and maintained towpaths along the river's banks to clear the way for tow teams making their way upstream. This was not without some undesirable environmental consequences. Without tree roots anchoring their soil, banks gradually eroded under the lateral force of the river's current, forcing its course to meander and branch into shallower side arms unsuitable for navigation.[8] In places where locals complied with requirements to clear trees along the banks, some simply threw the wood into the river, creating "more damage than advantage."[9] Any villages that lacked the means to keep their stretch of the river navigable faced grave consequences for their trade. As one travel guide declared of the small village Aschach near the Bavarian border, "The extensive sandbank which is yearly increasing before it, is an additional obstacle to its commerce, and Schultes [another travel guide] indulges in melancholy predictions respecting the ultimate fate of this unfortunate little town."[10]

For many other towns, however, the consistent delivery of raw materials and finished goods from the Danube contributed to thriving economic growth and transimperial trade.[11] In Upper Austria, salt sent down the Traun River arrived at Mauthausen on the Danube, where merchants transported it up to Bohemia.[12] In the late eighteenth century, one of the largest landowning families in the empire, the Schwarzenbergs, responded to lumber shortages by building a narrow canal and using several natural waterways to float wood from their heavily forested estates in southern Bohemia to the Danube and then downstream to Vienna. Manufacturers in Linz produced an array of iron goods, clothes, linen, thread, leather, and wool, which merchants "traded not only throughout the provinces and to the west but also sent down the Danube all the way to the orient." Even small towns like Marsbach and its surrounding villages housed wood workers whose "profitable trade" of goods was sent around Austria and "deep into Hungary."[13] In Lower Austria, inhabitants in Stein and its adjacent town Krems lived well from trade along the river. They welcomed ships and rafts coming down the Danube and Inn from Swabia, Bavaria, and Tirol, as well as ships rowing up from Hungary carrying wine, grains, and other wares. According to a contemporary source, Stein "serve[d] as the entrepôt for Moravia and Bohemia. Wine, wood, and fruit [were] all brought there *en masse* and shipped along on the Danube."[14] Vienna also served as a major port of commerce. In 1835, it unloaded more than 200,000 tons of fuel from the Danube, over 50,000 tons of building materials, and 12,000 tons of foodstuff, amounts that grew substantially when the population quadrupled in 1850–1910.[15] It also exported dozens of finished goods that merchants transported both up and downstream.

In Hungary, regional and international trade, an eighteenth-century population explosion, and the Habsburg reconquest of the Carpathian basin from the Ottomans had contributed to the growth of towns on the Danube. The population of Pest-Buda grew from approximately 50,000 at the end of the eighteenth century to over 120,000 by the 1830s, with about 8,000 barges "unload[ing] at the quay, in the course of the year."[16] Much of this trade specifically centered around the four annual fairs that Pest hosted, during which the Danube between Buda and Pest was for "half a mile in length ... covered with boats and barges."[17] Other cities on the Danube, such as Pozsony/Pressburg, Győr, Komárom, Zimony/Semlin, and Újvidék/Neusatz also hosted their own fairs and markets.

As a source of incoming goods and resources, rivers were attractive locations for businesses and residences. Housing advertisements declared that the location in question was "perfect for commerce on the Danube" or "close to the Austrian border, near the Danube" allowing a future owner to "profit greatly from any agricultural produce."[18] Contemporary observers credited Pozsony/Pressburg's favorable economic situation to its proximity to Vienna. The merchant town Baja also grew in importance thanks to the grain transports arriving in town and departing from its docks. By 1857, over 250 boaters and 53 wheat merchants existed in the town of 20,000 people; around 2,400 houses were used as storehouses for grain; and over 1 million carts came to the town every year for markets.[19]

While transportation was often easier by river than by road, trade was still subject to several restrictions. Municipal market rights (*Sta-*

pelrechte) forced merchants to unload and sell their wares in privileged cities before continuing their journey.[20] Local authorities also charged customs dues on wares arriving on the river. Some of these feudal and guild privileges fell during the enlightened reforms under Maria Theresa and Joseph II, but the others remained in place until the empire-wide customs union came into effect in 1850.[21]

The cost of transporting goods also fluctuated with a complex mix of factors. Széchenyi described how labor costs affected the profitability of bringing goods upstream: "Where areas are wilder and there is plenty of food for cattle to graze on . . . and where labor is cheap and humans idle, like on our Danube, then towing is profitable, as seen at market towns like Wieselburg [Moson], Becse, Pancsova, etc. But where there are landowners and no common grazing land along the river, then towing ships with cattle is expensive, and wherever laborers have something better to do than tow ships, then towing is also not worth it."[22] Changing weather and seasonal shifts in precipitation during the year also affected water levels and thus the reliability of river transportation. Insufficient river depth made transportation dangerous or impossible through unregulated and shallow sections. Too much rain increased the river's speed, driving up time and labor costs expended in towing goods upstream. Energy in the form of calories for cattle and supplies for laborers had to be balanced against the river's kinetic energy when deciding whether it was financially feasible to transport goods on the river.

Early modern river travel therefore provided many opportunities for shipping and traveling due to its ease, safety, and convenience, but there were limits to its profitability and reliability. Besides commercial policies and feudal privileges that restricted the free movement of goods and people, ships also literally ran aground on the hindrances that the river threw in their way; some of these formed by climate-induced erosion and deposition patterns, while others were a more unintentional result of anthropocentric activity.[23] Adaptations to the river were plentiful, though few changed the basic math of value, risk, and expense. As we shall see, steam engines and industrialization soon disrupted these calculations by reformulating the equation altogether.

Steam-Driven Ideology

In the late eighteenth and early nineteenth centuries, steam power technology and knowledge diffused throughout Europe and America.

This new technology encouraged speculation as to how it would transform interactions and relations between nations around the world. One British engineer wrote in 1825, "The progress of steam-navigation is truly grand and imposing: it is completely without a parallel in the history of art: it has facilitated the communication between the different states of Europe; and, by increasing the friendly and commercial intercourse of men in all countries it is to be hoped that it will soften prejudices, and remove sources of animosity, and contribute to ameliorate the condition of mankind."[24] The man who wrote these words, Thomas Tredgold, may have held an engineer's optimism, but his compatriots felt differently. Some pointed out that steam power would equip Britain's navy with an unassailable defensive strength and others that it would provide the means of spreading Christianity to all corners of the earth.[25] Neither exemplified the pacifism or goodwill that Tredgold envisioned.

In the early days of its introduction to the Habsburg state, steam navigation likewise sparked a spirit of optimism among contemporary observers, who saw steam navigation's expansion as a panacea for economic lethargy, social woes, and political tensions. As chapter 1 has detailed, members of the Habsburg dynasty hoped it would serve as a linchpin for commercial dominance on the Danube. Its appeal was apparent to others in the empire as well. As a young man István Széchenyi visited Great Britain. As a strong proponent of reform in Hungary, he marveled at Britain's advanced state and attributed its economic vitality and political liberalism to the widespread application of steam power.[26] Hoping to reverse Hungary's economic backwardness, Széchenyi bribed a British customs official and spirited a steam engine out of the country, seeking to use the tool to enliven transportation on Hungary's waterways. In the work Széchenyi published about Danube navigation, he stated, "Steam navigation can have a wonderful influence on the trade of our country, where there is no ebb and flow to worry about; the rivers here and there are narrow and rocky, but it is simply up to us to make this vision a reality."[27] Széchenyi's vision was not merely one of economic advantage. Rather, he viewed the new technology as freeing Hungary from the social and political conditions that remained from its feudal past.[28]

Like Széchenyi, the Irish journalist Michael J. Quinn imagined that steam navigation would transform Habsburg society. Arriving in the empire in the early 1830s when steam power was in its infancy, he embarked on a steamboat ride down the Danube. He noted approvingly that such technology would bolster communication and travel in the diverse empire, growing the region's wealth and development. He declared, "Steam navigation of the Danube will be a most powerful instrument of civilization: for it is quite true that steam and civilization are daily becoming almost convertible terms."[29] Besides his faith that steam power would provide positive social change, what Quinn implicitly acknowledged was the comparative backwardness of certain regions that needed to be "civilized" through steam power. In doing so, he was simply reflecting the social, economic, and even political hierarchies that ethnographers were formulating both within the Habsburg state and with other parts of eastern Europe.[30] Andreas Malm in his work *Fossil Capital* has critiqued the widespread, nineteenth-century opinion that steam power "civilized" humans, arguing that once steam engines were harnessed to produce goods, they tended to "accumulate" and reify positions of economic and social power at the expense of unindustrialized regions.[31] From the perspective of nineteenth-century travelers, however, traveling more broadly thanks to steamboats became the means by which they perceived and reified categories of "civilized" and "Oriental," particularly as they traveled down the Danube. As boats crossed from Habsburg into Ottoman territory in the 1830s and 1840s, passengers often remarked on the wild and undeveloped state of the vassal principalities Serbia, Wallachia, and Moldavia.[32]

With enthusiasm for steam navigation on the rise, the Danube Steam Navigation Company and the imperial government readily pursued efforts to expand offerings. While a single steamboat plied between Vienna and Pest once a month in 1830, by 1835 steamboats traveled from Vienna to the Black Sea once or twice a week and by 1841 trips on the river had increased sixfold. Twenty-two cities in the empire had steamboat agencies and landing places.[33] In 1838, the Vienna-based newspaper *Der Adler* rhetorically asked whether Austria was "progressing." It answered affirmatively, delineating the state's advances in technology, industry, trade, and population, and pointing out its business-friendly environment manifested by railways and "the Danube's coverage in steamboats."[34] This did not mean that steamboats dominated the river. In the subsequent decades, traditional shipping continued to rival steam navigation on the Danube. In 1835, the Danube Steam Navigation Company had six steamboats, whereas one of the wealthiest shipbuilders in the empire, Matthias Feldmüller, managed 350 ships regularly going upstream from Persenbeug to Regensburg and another 850 ships traveling downstream from Vienna to Pest-Buda.[35] One statistical report at the time suggested that the total number of ships and rafts on the Danube was likely between 6,000 and 7,000.[36] Thus, even with their superior ability to travel upstream, the total haulage of steamboats was a metaphorical drop in the bucket compared to conventional vessels.

The changes that steamboats brought to the river were seen in the nature and quantity of goods moving along the river. Requiring a steady supply of coal to power their new steamboats, the Danube Steam Navigation Company (DDSG) provided regular coal deliveries to towns along the Danube so docked boats could restock their supply. This ferrying of coal on the river was aided in 1838 when the DDSG launched its new tugboat *Erős*, the first mechanized towboat on the Danube, which brought coal from the Bánát in the empire's southeast borderlands to stations upstream.[37] To speed up the adoption of coal-powered shipping, István Széchenyi petitioned the imperial authorities to allow the DDSG to secure wood and coal from state lands.[38] By 1848, the Danube Steam Navigation Company's 50 steamers and 150 merchant ships made it the largest single coal consumer in the empire, more so than the empire's rail networks, which were the third longest in Europe after Great Britain and the German states.[39] Further bolstering the company's supply of coal, the DDSG bought mines at Pécs/Fünfkirchen (in south-central Hungary) and eventually built a mining colony for its workers complete with schools, grocery stores, hospitals, a church, restaurants, and housing.[40] These mines initially produced enough coal to cover the DDSG's own demands and it was even able to turn a profit from the sale of excess coal. To transport coal from its mines to the Danube, the DDSG built a freight rail connecting the mines at Pécs to the docks at Mohács.[41] Opening in 1857, the railway's popularity and importance for the region quickly expanded and soon began transporting passengers between Pécs and Mohács as well.

The growth in steam travel heralded a noticeable increase in trade and transportation between Austria and Hungary. From 1839 to 1847, the DDSG's general assembly quadrupled its Danube fleet from ten to over forty steamboats. The stretch from Vienna to Pest-Buda accounted for approximately 60 percent of the company's passenger and freight traffic and the frequency of trips on popular routes increased with ridership.⁴² As Hungary's reformers tried to liberalize the economy in the 1830s and 1840s, steamboats contributed to blooming tourism at Pest-Buda. Modern, luxurious hotels opened to welcome foreign visitors. The Angol Királynő (Queen of England), István Főherczeg (Crown Prince Stephen), and Európa Hotels became prominent destinations for travelers.⁴³ Companies published travel guides to help conduct passengers down the Danube and find accommodations and special sites for them in all the towns and cities they passed or stopped in. Communities wrote to the Danube Steam Navigation Company petitioning to be included on the company's burgeoning network of steamboat stations. Landowners in Paks, for example, wrote in 1846 to ask for a landing bridge "in the interests of the company and the public."⁴⁴ By mid-century from 1840 to 1868, the number of people traveling on the Upper Danube increased from 74,000 to over 2 million annually.⁴⁵

Business between the two regions boomed as well. Steamboats started running livestock ships on the Danube, a lucrative business that jumped from 9,000 head of cattle transported in 1839, one-tenth the number of cattle exported out of Hungary in that year, to 60,000 in 1847.⁴⁶ In general, the decade from 1830 to 1840 saw the value of agricultural products sent from Austria to Hungary more than double from 2 million to 4.5 million florins thanks largely to the value of animals for butchering (*Schlachtvieh*), "drinks," and grains. Agricultural goods sent from Hungary to Austria meanwhile dwarfed those figures, increasing from 16 million to 25 million florins, mostly thanks to the value of grains and vegetables, animals for butchering, and tobacco. The value of industrial goods sent from Austria to Hungary also doubled from nearly 18 million to over 38 million florins, mostly from fabricated items, woven material, and raw materials. The value of Hungary's "industrial" trade to Austria hovered between 25 million and 29 million florins and was characterized by "raw materials" (*Roh-Stoffe*), raw and partially processed materials, nonprecious metals, and sheaves.⁴⁷ In 1846, 13,000 tons of goods were kept at the local steamboat landing zone in Vienna specifically for shipment between Vienna and Hungary.⁴⁸ By 1847, the Danube Steam Navigation Company's landing place in Vienna held over 30,000 tons of goods connected to its trade with Hungary.⁴⁹

New policies and better steamboat designs made freight shipments more dependable and widespread. Previously, tugboats left landing zones once their cargo load was full rather than departing at scheduled intervals. A new policy in 1847 declared that boats would leave from assigned landing places on a fixed schedule. One author writing in a book about travel along the river stated that the result of this policy was "a new, massive flow of goods to the Danube."[50] More reliable shipping times encouraged businesses to integrate the river into their commercial ventures. Post carriages in Scheibbs, for example, coordinated with steamboat arrival times to offer rides from the Danube into the countryside.[51] The Hungarian Central Rail Company in Pest received parts, machines, equipment, and materials for its railcar production from Vienna sent via the Danube.[52] These goods and more were assiduously tracked by imperial sources and published in annual statistical reports.

By the late 1830s, better steamboat designs and stronger engines brought steamboats to more and more upstream communities. Steamboats were able to sail up the Danube from Vienna for the first time in 1837, extending steam freight and passenger traffic to important trade and manufacturing cities like Stein and Linz. In 1838, a steamboat sailed up the Sava River to Sziszek (Sisak, Croatia). While Sziszek's townspeople celebrated the arrival of their first steamboat, seasonal flooding and fluctuating water levels closed traffic, and the next ship did not arrive for another four years. In 1843, a new twenty-five-kilometer route sent boats up the Drava River to Eszék (Osijek, Croatia), Croatia-Slavonia's most populous city.[53] The Tisza also welcomed its first DDSG steamboat in 1833. In 1845, the DDSG negotiated with residents of Szeged to provide warehouse and harbor space to accommodate what the company anticipated would be a significant uptick in river traffic.[54] This bet paid off and soon boats not only called in Szeged but even traveled hundreds of kilometers up the Tisza to Tokaj.

These early years of steam navigation on the Danube took place against the background of tense political relations between the imperial authorities in Vienna and representatives of the Hungarian Diet. Hurting for foreign investment and worried about markets for their exports, members of the 1843–44 Hungarian Diet requested that authorities in Vienna abolish the tariffs on goods that Hungary exported to other parts of the empire.[55] Certain of Emperor Ferdinand's advisers were inclined to agree. Metternich and the head of finances, Baron Karl Friedrich Kübeck, plotted to remove the tariffs between Hungary and the rest of the empire to offset the empire's exclusion from the German *Zollverein* (customs union) after 1834. They also hoped it would moderate the influence of British trade. However, they were unable to overcome resistance from Magyar nationalists, aristocrats, and local county authorities who felt that such a centralizing and liberalizing policy threatened their historic political and economic privileges.[56]

Commercial and environmental conditions also challenged steam navigation in its first few decades. A trade crisis and plague outbreak in 1837 drastically dampened freight and passenger transportation, massive ice flows in winter 1838 damaged Danube communities, and October storms and low water levels in 1839 curtailed the season's duration earlier than usual.[57] The DDSG general assembly in 1838 acknowledged in its annual report that "nowadays, the regularity of our goods-transport arrivals remains entirely dependent on the river's mood."[58] But by the 1840s, trade in agricultural and finished goods again increased. Cash, horses, dogs, and even fortepianos joined the DDSG's expanding list of shippable goods.[59]

Taking a step back, what was not clear to contemporaries, who celebrated this early expansion of steam navigation throughout the empire, was the effect that it would have in both the medium term within the empire and in the long term in global history. Just looking at a few industries transformed along rivers in the Habsburg Empire: for Komárom's flourishing shipbuilding industry, its renowned galleys disappeared as steamboats ended the long raft trains, which had provided the city with abundant lumber imports.[60] As steam engines spread throughout the Habsburg Empire, demand for firewood as a fuel source diminished and was replaced by an insatiable thirst for coal. In this changing energy regime, some forests gradually regrew. Other forests were repurposed. Paper industries in the western provinces emerged in the early nineteenth century and continued to fell trees and use rivers for transportation, processing, and washing away effluence.[61] Likewise, the valley woods along the Tisza, pine forests near Komárom, and forests in Croatia-Slavonia previously used by shipbuilding industries were all henceforth available for other uses instead.[62]

This transition from a biomass to fossil-fueled economy was hardly unique to the empire, and countries around the world replicated it in various forms over the coming centuries.[63] On the other hand, this new fossil fuel age was a sharp break in global history. Those who saw the benefits could only marvel at the power that steam engines unleashed: mechanical, economic, and political. They did not worry about the carbon dioxide that such combustion released, which accumulated in the atmosphere for centuries thereafter.[64] While some scientists theorized about anthropocentric climate change, it focused primarily on desertification brought on by deforestation.[65] It would be up to future generations to take stock of the "advantages" of industrialization and weigh them against the unintentional and disastrous consequences that they wrought. Instead, other challenges would preoccupy nineteenth-century leaders in the Habsburg Empire.

The Danube as an Integrative Force

In 1848, in the midst of profitable and proliferating trade on the Danube, revolution broke out. To support Hungary's bid for independence, the newly convened *honvéd* (home guard) forces commandeered the ships of the Danube Steam Navigation Company. Serbian forces and imperial troops loyal to Emperor Ferdinand's regime did the same. In response, the emperor's troops closed the Danube to ships coming upstream from Hungary to safeguard the imperial capital at Vienna. Commerce stalled. After nearly a year of fighting, General Haynau of the imperial forces captured Komárom fortress, the Hungarians' critical defensive position upstream from Pest-Buda, which one visitor later called "quite the strongest fortress in the empire."[66] With Hungarian forces routed, this victory enabled the imperial authorities to reopen the Danube to navigation. Damages were tallied. The DDSG administration declared that closing the Danube to river traffic had devastated its bottom line due to the volume of business that it conducted in Hungary. Likewise, the Lower Austrian provincial government registered a large drop in its customs revenue the following year (600,000 florins less than in 1848), which it attributed to the Danube's closure.[67] Despite this short-term disruption, the following decades would see the Danube reemerge as an integrative force of trade and travel in the empire. How?

In the aftermath of these uprisings, the imperial government focused on reestablishing and growing empire-wide trade. The neoabsolutist authorities did so by overhauling economic and commercial policies on the Danube. New policies threw off the final feudal vestiges of privilege that had long hindered free trade and movement within the empire to forge a more open, imperial economy. First and foremost,

Trade Minister Bruck accomplished what his predecessors could not and integrated Hungary into the new empire-wide customs union, abolishing the dues and tariffs on goods transported between Hungary and the rest of the empire. The following year, the Finance Ministry abolished all ship fees on the Danube and its tributaries from Bavaria to Hungary.[68]

Both at the time of the customs union's announcement and for decades to come, politicians and scholars debated how well it actually served to integrate Hungary into the empire. John Komlos analyzes how Hungarian nationalists excoriated Hungary's heavily agrarian economy, which they blamed on the customs union and on the imperial government's treatment of Hungary as a grain-producing "colony." At the same time, he claims that others, including many Hungarians, considered the customs union to have an overall positive effect in Hungary, thanks to the benefits its economy received from access to Austrian credit and markets. Scholars continued this debate in the twentieth century. Hungarian scholars in the 1950s mimicked the rhetoric of nineteenth-century nationalists by characterizing Hungary's relationship in the empire as semicolonial. By the 1970s, however, most economic historians, Hungarian or not, had concluded that Hungary had profited from its closer integration into the imperial economy regardless of nationalists' grumblings at the time.[69]

Evidence suggests that the customs union alone did not cause the uptick in trade between Austria and Hungary but was one of several factors—including regulation works, population growth, and industrialization—that helped intraimperial trade take off in midcentury.[70] The effects of these developments were abundantly evident: in 1851–55, the DDSG's average annual freight traffic was five times higher than average annual figures in 1846–50, and enjoyed continual growth almost without exception until 1914.[71] By 1865, the DDSG had trafficked 487,000 tons of wares between Vienna, Pressburg, and Pest-Buda.[72] From 1861 to 1869 the DDSG shipments increased from 330,000 tons to over 715,000 tons, transporting things like lumber, salt, coal, iron wares, flour and tobacco. For decades after the 1867 Ausgleich, the DDSG's passenger and freight line with the highest traffic remained between Vienna and Pest.[73]

The Danube's extensive tributary network united different regions and communities in trade. In 1865, ships carried approximately 4 million tons of goods along the Danube and its tributaries. The largest amount of this freight, more than one-third, was loaded at communities along the Danube's tributaries and unloaded at communities along the Danube. Nearly a third of all goods traveled between different communities on the Danube and another quarter traveled between different communities on the Danube's tributaries. The rest, approximately 10 percent of all traffic, traveled from the Danube to communities along its tributaries.[74] Products like wheat, wood, and merchants' goods topped nearly 1.5 million tons on Hungary's stretch of the Danube and a whopping 2.5 million tons on the Danube's major tributaries.[75] One travel guide gushed that "thirty-one stations on the Upper Danube and seventy-two on the Lower Danube give evidence of the large traffic of travelers and goods that the company transports daily, and the importance of this undertaking is without parallel in the old world or new."[76]

To encourage domestic production and trade, the imperial government strategically liberalized commerce with some neighbors while raising protectionist tariffs against others.[77] After the 1856 Treaty of Paris internationalized the Danube and allowed vessels of foreign countries to freely sail on the Danube, the Habsburg Empire and other riparian states asserted an exclusive right to ply steamboats on the Danube. Great Britain, France, and other states condemned this decision. The Habsburg authorities also lowered tariffs to the west as a prerequisite to gain access to and exert influence in the German states' *Zollverein*. At the same time, it took measures to protect the empire's grain-producing regions against cheap, imported grain from the Ottoman Empire and the Danubian Principalities. The Finance Office, for example, requested that border customs inspect boats importing grain from the Sava River, and raise tariffs if necessary.[78] In 1857, Finance Minister Bruck extended this protectionist policy to several over imported wares arriving at the empire's border via the Sava River.

Protection against cheap imports was critical for many regions that sent their agricultural produce to the Danube to be distributed throughout the empire. This was particularly important in Hungary, where grain production grew heavily from the late 1840s onward linked to increased demand and higher prices.[79] Nearly 400,000 tons of grains flowed along Hungary's tributaries and 225,000 tons on the Danube in 1865 alone.[80] Even with rail lines growing exponentially in the 1850s thanks to the 1854 Railway Concession Law, which privatized the state railway and freed up private investment and construction, only one-third of the grain arriving in Pest-Buda came via rail. Traditional vessels, supported by steamboats, delivered the bulk of the city's grain supply.[81] At places like Pancsova in the Military Frontier, over 18,000 tons of grain left on steamboats and galleys—surpluses imported from the surrounding region that were then exported on the Danube.[82]

With new laws in place to stabilize trade, difficulties nevertheless arose to hamper river traffic. After the Railway Concession Law's passage, railways expanded from 125 to over 1,000 miles in Hungary and from 800 to nearly 2,000 miles in Austria. The ease of rail transport depressed passenger traffic on several main sections of the Danube.[83] In 1851, when a new rail line opened between Vienna and Pest, one of the DDSG's most trafficked stretches, its passenger numbers declined by one-third. The DDSG tried to reverse this trend by advertising lower prices and broadcasting whenever passenger traffic opened for the season in popular Hungarian newspapers like the *Pesti Napló* and *Budapesti Hírlap*.[84] Unfortunately, unseasonably low water levels and cold winter snaps that decade wreaked havoc on business, halting trips earlier in the season than usual.[85] Unsurprisingly, a cholera outbreak in 1855 also dampened people's desires to travel.

Despite these challenges, the groundwork for steam navigation's expansion was in place. The 1857 Danube Navigation Act gradually loosened the DDSG's monopoly and later boosted passenger and freight figures. After the 1867 Ausgleich, Hungary established its own ministry to oversee inland navigation, and the newly created Royal Ministry of Public Works and Transport (KMKM) received dozens of petitions from local businessmen requesting permission to set up steam navigation companies. Many cited the 1857 Danube Navigation Act as the basis for their petition. A few petitioners claimed that the imperial government had already validated their right to start new navigation companies in 1859 when the emperor's representative (*Statthalter*) in Hungary permitted a company in Pest to manufacture freight ships to compete with the DDSG. In a few short years, the Central Statistical Office in Budapest registered over fifty steam navigation companies of various sizes operating on the Danube, Tisza, Sava, and Drava Rivers.

This expansion came at a time when nationalist voices in Hungary clamored for the establishment of a *national* Hungarian steamboat company to rival what they perceived to be the dominance of the "Austrian" DDSG. The Central Danube Steam Navigation Company, the First Hungarian Steam Navigation Company, and the National Steam Navigation Company were all products of this era. Despite emphasizing their national appeal, each struggled to flourish. Bowing to financial pressure, they formed a single company, the United Hungarian Steam Navigation Company in 1871. Even this was unable to overcome the DDSG's supremacy, and the conglomerate eventually sold its ships and stocks to the DDSG in 1873 and folded, much the way the southern German Bavarian-Württemburg Company had done a decade earlier.

Instead, more successful efforts stemmed from smaller companies that served niche regional needs. Local petitioners seeking to establish steam navigation companies shied away from nationalist rhetoric, and cited benefits for their communities rather than nationalist antagonism as reasons for establishing their companies. In their petitions, local merchants and entrepreneurs emphasized that a decentralization rather than nationalization of steamboat commerce was preferable as it would better integrate local and regional business with traffic on the Danube. Rather than focusing on financial or regulatory hurdles, functionaries of smaller steamboat companies instead cited the Danube's ever-changing hydrological conditions and the need to guarantee navigation as the greatest determinant of their commercial success. The DDSG echoed these points, complaining to the Ministry of Public Works and Transport that ship mills and masses of tree trunks and branches continued to hinder navigation on the Maros, Samos, and Tisza Rivers.[86] While many smaller operations either went bankrupt or were bought up by larger companies later in the century, several regional steam navigation companies were founded in the 1870s, and they survived until the empire's entry into World War I.

By the last few decades of the century, steam transportation on both rails and rivers expanded significantly. The Budapest Statistical Office recorded a jump in new steamboat companies from 28 to 37, which operated around 1,200 ships between them.[87] From 1876 to 1896, rail transport to Budapest jumped from 17 million to 47 million tons per year, while waterways increased from 6.5 million to 14 million tons during this same period. While water freight represented a smaller proportion of overall goods, its growth rate was 75 percent compared to rail's 57 percent.[88] Austrian waterways likewise experienced a 65.6 percent rise in freight traffic in 1881–91, a growth rate higher than that of rails (62.8 percent).[89] One observer from Újvidék/Neusatz (modern Novi Sad, Serbia) wrote of the relationship between rail and river traffic: "One might believe that river traffic would completely decline after the railways' construction, but this view is refuted by the following data ... until the construction of the Budapest-Zimony railway, the traffic at the steamboat station—with exports—averaged 150,000 centners [9,260 tons] of cargo annually. Of that, [4,940 tons] was grain, [925 tons] was wine, and [1,850 tons] was flour. ... After 1883, the year the railway opened, until today [1894], the station has averaged [12,350 tons] of exports. Of that, [6,170 tons] is grains, [620 tons] is wine, and [3,700 tons] is flour."[90]

Such growth did not occur by happenstance but was the result of intentional cultivation. In Hungary the Royal Trade Ministry paired hydraulic engineering projects with quantitative assessments undertaken by the Royal Rail and River Inspectorate, which audited the traffic entering and exiting steamboat stations throughout Hungary in the early 1890s. Ministry officials then made recommendations about which stations remained viable to maintain and which had too little traffic to justify keeping open.[91]

In Austria, the dual onslaught of rail and steamboat travel drastically changed the composition of traditional watercrafts on the Danube. From 1858 to 1874, one River Engineering Office in Upper Austria calculated that of the nearly 127,000 vessels that had passed that stretch of the river, 30 percent were steamboats and the rest were traditional galleys and rafts.[92] From 1870 to 1890, rafts, galleys, and barges sailing down the Danube decreased 90 percent. Their decline was inextricably linked to the rise in steamboat traffic, which could more readily sail downstream and upstream against the Danube's current.[93] While draft animals, galleys, and rafts slowly disappeared, fruit growers, who floated their wares on narrow boats to downstream markets, began bargaining with steamboat captains to tow them and their empty boats back upstream.[94] Likewise, the DDSG continued to use hundreds of traditional vessels to serve local needs and travel up smaller rivers.[95]

At the end of the nineteenth century and in the early twentieth century, the Danube remained a critical avenue for shipping goods and people around the empire, serving to unite communities in Austria and Hungary. The Hungarians acknowledged as much during one of the decennial customs union negotiations wherein representatives from Vienna and Budapest discussed common imperial affairs such as the budget, external tariffs, and matters of the joint ministries (Finance, War, and Foreign) per the Ausgleich agreement. Negotiations were normally fraught with antagonism, but before the 1875 meeting, the Hungarian delegation published several positive statistics about the trade and relationship between both halves of the empire. While emphasizing the Danube Steam Navigation Company's prodigious growth figures between 1835 and 1875, one report stated, "It is common knowledge that a large part of this company's traffic is directly related to the traffic between Austria and Hungary, as the Danube is a natural partner of two parts of the empire."[96]

Both halves of the empire participated in this partnership. From 1886 to 1895 trade from Austria to Hungary doubled from 1 million to 2 million tons. A third of this trade came from combustible fuel: wood, coal, and peat. Other large percentages came from iron, drinks, and "minerals." Trade from Hungary to Austria increased from 2 million to 3 million tons. The largest shipments were grains, rice, fruit, vegetables, livestock, and other agricultural goods.[97] Everyday items like brushes, sieves, flaxen goods, rubber, and rosin saw their numbers tick up the fastest in trade between both halves of the empire.[98] More than half of the freight that the DDSG transported traveled from Austria to Hungary via its steamboats. Likewise, 85 percent of the freight that DDSG steamboats unloaded in Austrian towns originated from within the empire, and an overwhelming 85 percent of that domestic freight originated in Hungary. By 1907, over a million tons of river freight was exchanged between Austria and Hungary. This amounted to 15 percent of Hungary's freight shipments to Austria and more than 33 percent of the value of Austria's freight shipments to Hungary.[99] Broadly speaking, the interdependence of the two economies grew until World War I, with John Komlos calculating that Hungary exported 20 percent of its gross national product to Austria in 1913.[100]

This integration, or the appearance of DDSG dominance, did serve as a flashpoint for Hungarian nationalists. The minister of Hungarian Public Works and Transport (and later trade minister), Gábor Baross, implemented a series of policies in the mid-1880s that supported the expansion of rail transport. Part of these included a "transport tax" on water transportation, which the DDSG castigated as a blatant effort to support the new Hungarian State Railways and suppress the DDSG's river traffic.[101] Baross was also a strong proponent of regulating the

Iron Gates and establishing a national Hungarian steamboat company. He died in pursuit of the former (catching a cold while inspecting the rapids) and the Hungarian government eventually formed the Hungarian Royal River and Sea Joint-Stock Company in 1894 after his death. Others likewise took up the mantle of Hungarian nationalism along the Danube. In 1911, Count Tivadar Batthyány, the vice president of the newly reformed 1848 and Independence Party, led a spirited debate taking aim at Austria, given, in his words, "its pernicious efforts to repress Hungary." Batthyány directed his antipathy at the DDSG, arguing that the DDSG's subsidy agreement with the imperial government in Vienna meant that its commercial activities in Hungary were "in the service of Austria's economy." He concluded by exhorting the Hungarian government to do more to encourage Hungarian companies to haul freight on Hungary's abundant river system.[102] Despite broadsides from nationalist politicians and ministers, the DDSG remained a diminished but formidable force on Hungarian waterways.

While trade-minded officials, businesses, and producers generally applauded robust river traffic, it represented a grave threat in the minds of one segment of the populace: fishers. Catches had diminished throughout the nineteenth century and fishers' guilds and fishing clubs claimed that this was a result of systemic changes to the river in the nineteenth century. These groups blamed river regulation works and steamboat traffic, which they claimed disturbed fish life cycles, the former by reducing the suitably slow side channels ideal for spawning grounds and the latter by causing large waves that washed eggs and hatchlings up onto dry land.[103] Fishing associations in Austria vocally backed their Hungarian counterparts.[104]

Contrary to these claims, fishing communities had been buffeted by a more complex array of forces for decades. With the expansion of farmland to feed the empire's growing population and the establishment of factories along the Danube, both irrigation and industrial processes had begun exploiting water resources and polluting waterways. It was not until 1869 that a new law in the Austrian half of the empire was passed governing the common and public usage of water.[105] A sharp rise in food prices in the late 1870s due to crop failures also led to a concerning rise in illegal fishing. Likewise, railways, refrigerator trains, and higher meat prices all increased the demand for and accessibility of fish, depleting their stocks from the empire's rivers.[106]

In the 1870s, dozens of fishing associations and clubs banded together across the empire to protect their livelihoods. Some tried to mitigate the effects of steam navigation. At an 1884 fishing conference in Vienna, one speaker tried to persuade participants that fishing communities would always feel tension with steamboat companies, so they should stop trying to fight them and instead attempt to negotiate compensation for fishers' losses and suggest compromises that would help fishing communities.[107] In addition, many embraced the new scientific optimism around pisciculture—something that earned the attention of the imperial and national authorities. These groups effectively petitioned for financial backing and subsidies by explaining that increasing fish stocks supported general well-being by moderating food prices and increasing the economic value of the industry.[108]

It was clear that government agencies, representatives, and businesses seemed to prioritize commerce over catches. The diet in Budapest passed Law 19 "On Fishing" in 1888 to address declining fish stocks in Hungary's waterways. Decades of scientific studies launched by a joint Austro-Hungarian commission had concluded that overfishing and indiscriminate practices, like dynamic fishing, were one of the main culprits for the drastic decrease in fish stocks. The new law banned fishing behavior that was deemed exploitative or excessively wasteful. While experts agreed that steamboats and regulation works also played a role in diminishing fish populations, one sympathetic article summed up the general attitude: "Although [these projects] have a detrimental effect on fish reproduction, [they] benefit the public interest to a disproportionately greater extent in other directions."[109] Thus, halting them was not an option. Fishing communities had to adapt to the new realty of steam navigation, or they would decline as certainly as their catches.

Steam navigation was, quite simply, too important for transimperial trade. By the early 1900s, several steamboat companies were on the Danube trafficking between both halves of the empire. These included the DDSG, the South German Steam Navigation Company, the Royal Serbian Steam Navigation Company, and two Hungarian companies. In 1908–10, the two Hungarian companies transported approximately 20 percent of all freight traffic on Austrian waterways while the DDSG carried approximately 50 percent.[110] In Hungary, the DDSG transported over 50 percent of all freight traffic on its waterways in 1900, gradually diminishing its stake to 40 percent by 1914.[111] Studying the DDSG's freight traffic reveals that an unmistakably strong commercial link remained between Austria and Hungary. Of the nearly half a million tons of goods departing from Austrian towns on DDSG ships in 1913, 58 percent were destined for Hungary. Of the goods the DDSG unloaded in Austrian towns, 66 percent came from Hungary.[112] In 1910–14, the Hungarian River and Sea Navigation Joint-Stock Company exported an annual average of nearly 200,000 tons from Hungary to Austria (approximately 90 percent of its upstream trade), and imported more than 81,600 tons (approximately 90 percent of its downstream trade) from Austria to Hungary.[113] From these figures, the national Hungarian company's business did not seem adversely affected by doing business with Austria, which accounted for the lion's share of its "foreign" imports and exports up to World War I. Indeed, just as rafts, barges, and other crafts had transported goods and people between Austria and Hungary for hundreds of years, steamboats became the latest manifestation of this Danube traffic, integrating people and communities around the empire.

Cities, the Drivers of the River Network

What accounts for the growing trade and the diversifying products that ships transported along the Danube during this period? Growing populations and growing cities. By the 1910 census, almost 20 percent of the empire's population lived in cities of over 100,000 people, twice the number from 1869.[114] Four of the empire's twenty largest cities were located on the Danube and nine other major metropoles were located on one of its tributaries.[115] These growing cities needed increased access to resources, such as lumber, coal, food, construction supplies, raw materials for industry, and everyday items for furnishing housing and businesses.[116] While chapter 5 will discuss the ways that urban residents engineered the river for business, recreation, and commerce, next we consider how the growth in riparian communities expanded connections along the Danube and its tributaries.

As cities grew in population after mid-century, thousands of vessels continued to supply them with foodstuffs, raw materials, and goods. Of the 31 towns in Upper and Lower Austria with steamboat stations on the Danube, nearly all imported more grain than they exported. In 1861, Vienna measured more than 477,000 tons of goods arriving and departing. By 1868, new facilities for loading and unloading freight helped freight traffic nearly double to 895,000 tons.[117] One travel guide attributed this rise in traffic to the DDSG's business strategy, pointing out that "since the DDSG introduced additional steam connections between Vienna and Pressburg with the goal of provisioning the capital, traffic and trade seem to have doubled."[118] In 1870, traditional vessels schlepped 532,000 tons of freight in and out of Vienna, and steamboats carried another 361,000 tons to and from the capital. Ledgers indicated that steamboats mainly transported merchants' goods and grains, while galleys brought construction lumber, firewood, construction stone, hydraulic chalk, bricks, coal, sand and gravel, foodstuffs, and clay.

Vienna imported about seven times as many commodities and goods by weight as it exported, a demand that influenced river-wide freight trends. In 1865 and 1866, for example, Vienna had a lull in construction projects, which caused construction lumber shipments on the Danube to drop by half.[119] The importance of supplying the imperial capital also kept many smaller ships afloat. While traditional vessels, like barges and rafts, had gradually disappeared more generally along the Danube, their numbers at Vienna increased. From 1887 to 1894, the number of rafts and galleys arriving in Vienna increased from 2,600 to 4,200. Traditional vessels also increased the tonnage they delivered to the capital each year until 1914, even as the proportion of freight delivered via steamboat rose from 40 percent to 70 percent from 1868 to 1910.[120] Vienna, nearing one million residents in the early 1870s, needed grain to feed its growing urban population. In 1870, approximately three-quarters of the goods reaching Vienna did so via the Danube.[121] In 1910, nearly all the grain consumed by Vienna's residents arrived via the Danube.[122]

While larger, growing cities, like Vienna, imported far more food, goods, and materials than they exported, about four times as much, smaller communities were integrated into the river economy by serving as entrepôts for burgeoning river traffic. With fewer than 6,000 inhabitants, Titel, a town between Pest-Buda and the southeast border, trafficked more grain than any other city on the Danube in Hungary except Pest-Buda and Győr. At another city, Baja, in the town of 20,000 people, nearly 2,500 houses there were used as grain storehouses. Over 1 million carts came to the town ever year for markets, and it was considered a "final destination" on the Danube for many regional businesses.[123] Farmlands around Baja transported well over 120,000 tons of grain a year to the city by 1860, more than half of which was transported by boat to larger cities along the Danube. One observer described how "in the spring and fall, flotillas of ships on the Danube and its branches swarm the city [Baja], while warehouses line the main streets specifically for these periods."[124] The movement of goods in these trade towns provided a crucial livelihood for residents.[125]

A small snapshot reveals the countless people in both large and small communities who were connected to this Danube economy. In 1868, the village of Gmunden on the Traun River sent nearly 450 tons of salt down the river to Vienna on galleys; Kupfstein on the Inn in the western province of Tirol sent nearly 850 tons of cheese, butter, lard, and eggs, and its upstream neighbor Schwaz sent 1,250 tons of iron and iron goods.[126] The DDSG's coal transport also averaged around 220,000 tons per year and the company conveyed hundreds of thousands of livestock annually.[127] Scores of other products flowed on the river, such as fruits, vegetables, chalk, construction lumber, granite, and many other things, demonstrating how diverse swaths of the population from salt miners to dairy farmers relied heavily on the river for their businesses and livelihood.

This riverine economy remained vibrant throughout the century, and municipal residents deployed great energy and ingenuity maintaining and expanding access to river traffic. When residents of Szeged petitioned the Ministry of Public Works and Transport for the right to establish a steamboat company in 1867, they advanced a vision of accelerated trade up the Tisza to "connect the city's economy more firmly to the Danube." At the junction of the Tisza, Danube, Maros, and Körös confluences, Szeged was, according to the petition, better suited than Pest or Vienna to integrate the markets of Bács, Csongrád, and Csanád Counties as well as those of the Bánát. As a final point, the petition emphasized how Szeged housed the largest boat factory in Hungary and that every year it welcomed a quarter of all private boats and ships on the Danube, Tisza, Drava, Maros, Körös, and Kupa Rivers.[128] After Szeged's mayor followed up with several ministries regarding the petition, they granted the city rights to establish a local steam company. Within two weeks of selling stocks, the new company raised enough capital to purchase two steamers, set up stations for coal, and pay the crews' salaries for both ships. Szeged's economy remained tied to its new business. Three decades after the company's launch, Szeged registered almost half its freight goods departing the city by river (43 percent departing on rails and 11 percent on wagons). An impressive 70 percent of freight traffic arrived in town by boat.¹²⁹ The Szeged Steam Navigation Company was not unique in these accomplishments. Local companies from Győr to Nagybecskerek looked to provide passenger and freight traffic in small segments of the Danube or its tributaries. Dozens of smaller companies even served single stretches of passenger or freight traffic.

For cities and communities without their own steam companies, petitions sought to maintain existing services on the Danube to serve "the general good," even as newly built rails threatened companies' profit margins. In 1880, of the 13 million passengers who traveled in Hungary, approximately 80 percent traveled by rail and 20 percent by river. When Trade Minister Baross introduced zonal pricing on rails in 1889, 10 million rail passengers jumped to 30 million the following year. River ridership remained steady at 3 million people. Confronting rail competition, the DDSG acknowledged that passenger lines represented a drain on its resources.¹³⁰ By the mid-1880s, the company's passenger growth had slowed. Reaching the apogee of its passenger traffic in 1890—3.5 million—it witnessed a precipitous halving of passengers by 1901. To keep itself afloat financially, the company sought to shift emphasis to its freight enterprise and cut some of its less profitable passenger stretches.¹³¹

Local communities fought against this potential loss in service. In April 1892, the Upper Austrian provincial diet (Landtag) drafted a plea to the imperial authorities in Vienna to pressure the DDSG to increase its passenger traffic on the Upper Austrian Danube, particularly for local communities from Engelhartszell down to Grein.¹³² In response, the DDSG instated a twice-weekly freight transport between Linz and Passau, which allowed passengers to buy tickets to ride on the deck. This provided residents with some limited transport options. The DDSG also offered the same service on its Vienna–Linz freight line. In 1893, it even arranged a local ship designated to carry passenger traffic between Stein in Lower Austria and Linz and Aschach in Upper Austria. In each of the subsequent years, the Upper Austrian provincial diet also appealed to the imperial authorities asking them to pressure the DDSG to guarantee certain local passenger services. In 1896, it called for the company to guarantee local passenger service to small towns along the Danube starting at the opening of the steamboat season, a time when traditionally only post ships were on the river, until wintry conditions ended the season. The DDSG, despite "unfavorable results" on local trips from Passau to Krems, complied "upon request from the high government and in the interests of the local residents."[133]

At the turn of the century, to improve passenger traffic on the Danube, the Upper and Lower Austrian Landtage discussed possibly linking river traffic with the burgeoning Promotion of Foreign Traffic (*Förderung des Fremdenverkehres*) movements in the different provinces, which had existed since 1884.[134] These movements proposed boosting economic development by attracting tourists, though certain agrarian interests voiced strong opposition to them. As local groups founded associations in cities and provinces to support this goal, they distributed brochures in German, English, and French and petitioned the imperial authorities to subsidize and advertise their activities.[135]

Cities tried to (re)capitalize on this movement to make communities on the river attractive for tourism. Melk's city council also tried to promote Danube traffic in Lower Austria, debating how to harmonize train and steamboat schedules to maximize the efficiency of transporting people to the Wachau from up- or downstream. Council discussions recognized that local practices, such as local businesses' opening hours, affected tourism to the area. The following spring, certain council members argued that "hospitality locations" should be permitted to stay open later—until 4:00 a.m. or 5:00 a.m.—to cater to early or late arrivals to town. The council also granted a local request to appeal to the DDSG for more convenient passenger traffic.[136] The Tourist Club in Melk also sent petitions to both the local rail company and the DDSG to request that the two companies coordinate their departure times, which would preferably connect to the community twice a day.[137] As tourism blossomed in the Wachau, Melk's city council requested funds to build a ferry to cross the river, claiming that the Danube was a source of "great patriotism," and the ferry would promote foreign traffic and better take advantage of the "imperial river." Advocates pointed out, for example, that tourists from Vienna, such as singing groups (*Gesangvereine*) liked to take excursions out to the Wachau. Provincial authorities granted the city's request, a decision that coincided with the arrival in town of the Lower Austrian viceroy Count Kielmannsegg, who came to Melk with ambassadors and delegates from Britain, France, Italy, Romania, Saxony, and Berlin to visit the Benedictine abbey and dine aboard a luxurious steamboat docked nearby.[138]

Tourism and spectacles in cities along the Danube were, overall, beneficial for steam companies' bottom line—a point the DDSG frequently exploited. In its 1867 business report, it recorded the positive effects that the Paris World Fair and Franz Joseph and Sisi's 1867 coronation in Buda had had on passenger traffic. In both instances, travelers had used the occasion of *other* events to incidentally enjoy Danube cruises.[139] When Vienna hosted the World's Fair in 1873, the DDSG had also attracted attendees by reducing prices for tickets down the Danube by 50 percent as well as reducing freight costs for objects and animals heading to Vienna for the fair. Unfortunately, while the DDSG had prepared ships and stations around Vienna for pleasure cruises in town and around the fairgrounds at Prater at lowered prices to accommodate the expected influx of visitors, poor weather and a subsequent financial crisis depressed turnout and disrupted general traffic on the Danube. The company glumly operated at a loss that year.

Tourism, world's fairs, and national celebrations kept traffic flowing to Hungary, where the public was treated to displays of innovations and progress. Especially after the 1867 Paris World's Fair and even somewhat after the 1873 flop in Vienna, local initiatives endeavored to bring attention to Budapest as well, with its modern features emerging from the city renewal projects and Danube regulation commencing in 1871.[140] In 1876, the press described the large delegation of visitors departing Budapest on DDSG ships to sail down the Danube to Constantinople for the International Congress of Ancient History.[141] In 1885, organizers in Budapest hosted the Countrywide Exhibition (Országos Kiállítás), which presented products from around Hungary meant to attract foreign visitors. Budapest hosted the World's Fair in 1896, which coincided with its Millennial Exhibition. The DDSG declared that business had boomed in 1896, thanks to both favorable river conditions and lively passenger traffic to and around Budapest related to these celebrations.[142]

In the lulls between major celebrations, tensions arose between citizens and the DDSG when it occasionally sought to pare back its less profitable river routes. In these instances, the public petitioned imperial and provincial authorities to intervene and compel the DDSG to both maintain and even expand passenger routes. These petitions detailed frustrations that local passenger transports were inconvenient or nonexistent, if ships arrived at all. Many complained about the high price of tickets. In Upper Austria, the provincial legislature turned to the authorities in Vienna and demanded that it require the DDSG to implement several measures to improve safety and convenience for passengers. Requests included installing telephone or telegraph lines between stations—so they could communicate with each other to notify travelers about possible delays with steamboats—and offering at least one daily passenger service traveling in both directions on the Danube. Should the DDSG not honor these requests, the diet members argued that the imperial government should shoulder the costs to guarantee navigation through other means.[143]

These efforts at the provincial level reflected the strong civic engagement at the municipal and associational level to maintain the network of river traffic for cities' benefits. In January 1902, the small municipality of Aggsbach in Lower Austria banded together with eighteen other communities to request a steamboat station from the DDSG. Their petition also arrived at the Imperial-Royal Trade Ministry, and expressed hope that a new station would continue supporting efforts to bring tourists to the region and connect their hinterland to the river.[144] Later that April, the "Upper Austrian Association for Foreign Traffic" (*Landesverband für Fremdenverkehr in Oberösterreich*) suggested ways that the Imperial-Royal Trade Ministry could help coordinate rail and ship schedules on the Danube for more favorable passenger traffic.[145] Additional petitions took up this thread, trying to determine how the company could better supplement arriving and departing trains or ensure that its ships trafficked at more favorable hours to avoid inconveniencing passengers with late night arrival times. The DDSG responded by printing pocket schedules to give travelers easier access to the company's steam schedules. It also maintained certain stretches "despite their unprofitability." The company also acknowledged that a particular petition it had received for a steamboat station at Akoven seemed worthy of consideration. This petition had effectively argued that agricultural producers in the region needed access to markets, and a steamboat station would solve the problem that they were "8 kilometers from the nearest rail station, 11.4 kilometers from the nearest steamboat station at Brandstatt, and 22.8 kilometers from Linz."[146]

Local communities did not limit demands to ship schedules. In 1904, the Upper Austrian legislature again acted to induce the DDSG to attend to local economic needs due to the "prominent economic significance of the Danube as a traffic way." These needs included expansion in the number and quality of landing stations, connection of telephone lines between stations, and the company's support in pres-

The Danube as a People Network

suring Hungary to remove its "transport tax." The provincial body presumed that the company would be more amenable to granting its petition in light of its impending contract renewal with the imperial government.[147]

At the end of the nineteenth century and particularly in the first decades of the twentieth century, persistent local initiatives, along with more favorable navigational conditions from regulated Danube stretches, led to a renaissance in traffic on the empire's waterways. Passenger traffic in Hungary, which in 1901 had reached its lowest point in decades with only 5 million passengers, nearly doubled to 9.2 million by 1912.[148] River traffic in Austria also rose. From 1902 to 1912, there was a 72 percent increase in traffic on the Danube between Passau and Theben/Dévény, and Vienna itself experienced a 109 percent increase.[149] This growth was particularly impressive considering the impasse that imperial authorities and the DDSG had found themselves in during their 1907 subsidy negotiations, which had led the company to threaten major cuts to its passenger services. Appeals from citizens and regional governments attempted to stave off this "disastrous" possibility.[150] Freight traffic also grew considerably in Hungary thanks to a variety of regional companies.[151] Behind these numbers were the complex local, provincial, national, and imperial deliberations and negotiations on how best to foster both steamboat and general traffic on the Danube.

Looking back at the nineteenth and early twentieth centuries, communities relied on the freight and passenger traffic proliferating on the Danube. Expanding urban populations required ever greater deliveries of food, fuel, and raw materials. Securing and transporting these goods fostered commercial linkages across the Habsburg Empire. Market mechanisms were not alone in forging these bonds. Individuals, companies, and representative bodies in riverine communities also appealed, petitioned, and coordinated with one another to nurture this expansive network of trade and transportation. Looking at petitions from various regions, people expressed their desire to be integrated into this river network with strikingly similar rhetoric, espousing hope that it would promote local economic interests and be in the greater interests of the empire. Imperial authorities were often inclined to agree. Thus, through trade, travel, and tourism, these communities benefited from burgeoning traffic, while serving as critical drivers of Danube River traffic.

The Empire as a Connected Community

Writing his German travel book *Illustrated Guide on the Danube from Regensburg to Sulina,* Alexander F. Heksch argued that "the Danube is not only a trade and traffic path but also has high importance in its cultural-historical relationship. The history and development of many nations in middle and south Europe are tightly bound with this mighty river.... Trade and ship traffic brought all these people together."[152] The Industrial Revolution strengthened and expanded such connective networks.[153] Global trade increased tenfold between 1850 and 1900 thanks to fossil fuels and new modes of transportation. The rise in steam power and the empire's energy transition from biomass (wood/human labor) to fossil fuel (coal) created new conditions along the river, and some promised to integrate disparate elements of the imperial riverine economy more closely. Energy resources from coal to foodstuff were increasingly shuttled about by the empire's growing steamboat fleets.[154] The Danube provided the main path along which finished goods and raw materials traveled by river, encouraging commercial and agricultural actors from its hinterlands and along its tributaries to form supply networks. This integrated traffic as diverse as the salt mines of Transylvania and the mountain pastures of Tirol.

For an empire feeling the strains of diverse populations and frequently competing ideologies, advances in transportation and communication transformed citizens' relationship with the Danube and with each other. Habsburg monarchs, the imperial bureaucracy, commercial groups, and eventually communities on the river used the new technology to pursue goals at the imperial and local levels. After the 1867 Compromise, in contrast to divisive political and national rhetoric in parliamentary circles, regional steamboat companies touted their success based on the integrative role that they served along the empire's river network. While unregulated river stretches challenged a linear progression, new cooperative relationships between steam navigation companies, commercial groups, and even railway companies forged new connections for citizens and communities with the river and with each other. These exchanges and connections revealed that despite the antagonistic and nationalistic rhetoric found in certain political circles, commercial and personal traffic on the Danube continued to strengthen bonds between the Austrian and Hungarian halves of the empire.

The Danube as a People Network

In the Habsburg Empire's final years, the Danube experienced a revitalization, continuing to serve as a transnational, intramonarchical network and to grow the connections between peoples, goods, and services. New and old technology combined to bolster the river's utility as river traffic expanded, thanks to growing the function of cities as producers and consumers and as destinations for travelers. War, flooding, trade interests, urban growth, international relations, and a desire to travel all contributed to both changes to the river and people's behavior along it, as the Danube strengthened imperial connections of goods and people around the empire. The war cut short these relationships, and the empire's successor states soon discovered the painful realities of independence and separation from the beneficial, socioeconomic space that the empire and the Danube within it had provided.[155]

Chapter 4

OVERCOMING DANUBIAN DANGERS

On August 18, 1849, the small community of Windisch-Matrei in the western Habsburg province of Tirol started the day festively, celebrating the birthday of the nineteen-year-old Habsburg emperor Franz Joseph. That night, however, residents were brutally awakened by an onslaught of flooding from the local mountain stream, the Bretterwandbach. As inhabitants scurried to save themselves and their property, a sense of *déjà vu* likely pervaded the mood. As the district leader later wrote, the community's finances were completely depleted after centuries of rebuilding dams, bridges, and property after such disasters.[1] This suffering was felt widely in the Habsburg Empire as floods had also struck communities along the Danube in 1830, 1838, 1847, and earlier that year in 1849.[2]

A few days after flooding subsided, a self-proclaimed "cameralist" wrote to the *Innsbrucker Zeitung* to discuss financial responsibility. The author provided as an example of financial *irresponsibility* the insufficient funding for projects like river regulation and flood prevention measures, claiming that floods resulted in expensive reconstruction efforts and onerous burdens on the population.³ The article asserted that there were shared responsibilities between governments and private interests to fund public works. Admittedly, this was hardly a radical new idea and had, in fact, circulated for decades in political halls from Vienna to Washington, DC.

The reason that such ideas held sway and gained prominence was that natural disasters exerted immense influence over global affairs and the cost of these disasters was painfully high throughout history. From 1300 to 1900 during the so-called Little Ice Age, unpredictable weather patterns in Europe reduced the Norses' Atlantic crossings to Greenland, inundated North Sea communities with floodwaters, threatened villages with advancing Alpine glaciers, and caused widespread famine on countless occasions because of heavy rainfall, drought, and crop disease.⁴ At the height of the Little Ice Age in the seventeenth century, pressures from these climatic variations had worldwide consequences; Manchu invaders successfully overthrew the Ming dynasty and established the Qing Empire, the Ottoman Empire underwent several decades of succession crises and deposed sultans and saw its European territory diminish, and Europe itself experienced ninety-seven years of warfare during the century, including the deadly Thirty Years' War. Despite social, economic, and political upheaval, notable exceptions like Mughal India, Safavid Iran, Tokugawa Japan, and a few Italian holdings of the Spanish Habsburgs demonstrated that rulers could weather climate crises (and rebellions) with well-conceived public assistance programs, ruthlessness, and political compromises.⁵ But vulnerability to climate shifts remained. In the early nineteenth century, four major volcanic eruptions and reduced solar activity lowered global mean temperatures, resulting in increased cloud coverage, precipitation, and glacial advances. These variations disrupted crop yields, contributed to greater flooding, and led to miserable conditions of hunger, high grain prices, and more frequent disease outbreaks.⁶

Floods frequently struck the river-rich Habsburg Empire. Naturalists had long understood the basic mechanics behind the hydrological cycle and flooding, with many observers assiduously measuring rainfall and charting the corresponding rise in river levels.⁷ Nevertheless, floods remained bafflingly complex phenomena with reinforcing natural and human-induced causes. Seasonal factors, such as spring rains, glacial thaws, ice floes, and summer rainfall influenced water volume and levels.[8] Geographical features from low-lying terrain to tree coverage also affected the distinct hydrology of the empire's rivers. For example, describing the local geography in Vienna, one commentator analyzed the causes of Danube floods, stating, "If one studies the Danube's conditions at Vienna, one notices how the water flows over a higher, broad riverbed, branching into several arms, and how its many meanderings hem its progress and cause it to lose speed as it breaks against many islands, banks, and sandbanks." The author went on to point out that these hindrances that slowed the Danube also caused it to flood, as it could not drain fast enough when heavy rain fell.[9] Moreover, cooler winters from 1500 to 1800 in Central Europe sharpened flood conditions, as unregulated rivers froze, then thawed, then formed ice floes and ice dams, which blocked rivers and forced water to overtop its banks.[10] Rainier weather between 1768 and 1789 also caused a higher than average number of floods on the Danube.[11]

Inhabitants in the empire were in part responsible for increasing the devastation of flooding. With the population in Hungary growing from four million to seventeen million from the eighteenth to nineteenth centuries, farmers and companies drained wetlands and cut down forests to plant crops.[12] These activities decreased absorption of rainwater and increased erosion along waterways.[13] Commercial activity, such as cutting down trees to clear towpaths and anchoring boat mills, also increased erosion, causing rivers to meander, become shallower, and flood more easily.[14] Even efforts to stave off flooding by raising levees made downstream flooding worse.[15] In these ways, residents contributed intentionally and unintentionally to not-so-natural disasters.

With debates raging in the government and in the public sphere about the shared responsibility for preventing floods, flooding in mid-century acted as a catalyst for political action. The indiscriminate threat that floods posed to the empire's diverse population, and the transregional destruction that they caused, provided space to debate and shape expectations about how exactly the government and citizens alike could help mitigate these disasters. Successive floods in the early nineteenth century demonstrated the inadequacy of the empire's disjointed and decentralized approach to flood preventative measures. In response, the imperial government enacted structural reforms to bolster its responsiveness to citizens' discontent. In doing so, it hoped to demonstrate its indispensability when addressing the complex and multifaceted challenges that inhabitants in the empire faced from nature. The residents of Windisch-Matrei celebrating Franz Joseph's birthday may have felt the pain of loss in the 1849 flood, but the emperor and his government rewarded them and many such local communities, as they became the prime beneficiaries of the new arrangements in flood prevention and relief efforts in the coming years.

An Unregulated Empire Before 1849

Before the mid-nineteenth century, large floods persisted in the Habsburg Empire because disparate local authorities remained responsible for funding and undertaking preventative measures. Early technical, commercial, and political bodies that organized flood protection measures in the Habsburg Empire tended to be particularistic, focusing on securing local stretches under their political jurisdiction or covered by their financial means. In the early nineteenth century, each province in the western half of the empire had its own separate engineering department, which undertook a variety of projects from regulating rivers to constructing embankments. Hydraulic works were frequently uncoordinated across departments, even along major rivers like the Danube that flowed through several provinces.

To fund public works, provincial engineering directorates introduced price hikes on commodities like wood, iron, towing rates, rope, and illumination and lubrication oils, which they announced in local papers, along with upcoming projects. Such advertisements justified the cost by delineating which public sites would benefit from the material and construction costs. The directorates also informed the public when and where the work might affect businesses or residences near the river. For example, one article publicized that due to damages to the "Danube Bridge" from recent "natural disasters," there would be a temporary ban on wine carts crossing the bridge until the directorate could stabilize it.[16] Such announcements covered various geographic regions, for example, informing readers about decrees mandating the construction of flood protection works near Linz in Upper Austria. A series of announcements in April 1831 appeared in the *Wiener Zeitung* and *Brünner Zeitung der k.k. priv. mähr. Lehenbank* broadcasting the imminent construction of embankments near Vienna, detailing specific houses and industries near which it would be working. Such work was not minute. In the decade from 1830 to 1841, the Viennese authorities constructed about fifty-five miles of embankments, a jump from the twenty-five miles of embankments they had constructed each decade from 1750 to 1780 during the worst flooding of the Little Ice Age.[17]

Policies in the empire's eastern half differed from those in the west. The National Diet in Buda governed territory along the Middle Danube, but government funds for public works were limited, so the Diet required local landowners and associations to organize their own embankment construction and floodplain reclamation. The 1807 Law XVII mandating that landowners undertake their own improvements and protections also stipulated that they could obtain money from the state if they did not receive full support from other stakeholders (the latter could not receive profits derived from the enterprise).[18] Some cities and counties organized efforts, such as a 1700 law in Pozsony/Pressburg requiring both serfs and nobles to contribute to embankment construction, though such work rarely extended beyond municipal boundaries. One technical solution to avoid flooding involved cutting ad hoc transections to eliminate the river's meandering path, a policy abandoned in the eighteenth century when it did not produce the desired results.[19]

Large landholding, aristocratic families like the Esterhazys, Festetichs, and Zichys were exempt from taxation and thus did not directly fund public works. In the eighteenth and nineteenth centuries, however, these families were some of the first to initiate projects like draining low-lying swamps and reclaiming alluvial plains to protect their land. Several nobles even supplied land for the twenty-mile-long Hanság Canal, which drained land around the Leitha and Hanság regions.[20] Drainage and reclamation did not stem from paternal benevolence but mostly provided aristocrats with tangible economic benefits. Protecting pastures and sheep protected wool, a valuable commodity in Europe-wide trade. Once grain prices rose during the Napoleonic War, nobles drained standing water from their holdings to increase arable land, crop yields, and profits.[21] However, these private interests provided a good foundation for later commercial endeavors in the 1830s erecting dams along the river.

Because residents erected barriers only to prevent *local* flooding, this uncoordinated approach failed to permanently or even consistently defend communities from floods. The Viennese authorities undertook intensive regulation and protection measures near the imperial residence city in the late 1780s and 1790s, which failed to fundamentally change the river's natural state.[22] Massive flooding in 1783 and 1789 caused so much damage to Danube communities in the Szigetköz, a large island downstream from Pressburg/Pozsony, that the residents from two counties raised funds to build protective dikes. Unfortunately, they were "so primitively built," according to a late nineteenth-century engineer, that a flood in 1809 destroyed them.[23] A 1775 flood at Budapest caused renewed efforts to build embankments, but a 1799 flood also destroyed them.

Attempts to coordinate actions and overcome this local particularism remained futile. Periods of interrupted or abrogated National Diets in eighteenth-century Hungary meant that counties became more protective of their local rights, thus calcifying political particularism. When cities in Pozsony County built embankments in the eighteenth century, flooding in downstream Komárom County worsened. In 1801, Sopron and Vas County officials tried to coordinate actions to regulate the Rába River, a prominent tributary starting in Austria and flowing into the Danube at Győr in Hungary, to no avail. In 1816, a Royal Commission brought together officials from Sopron, Moson, Győr, Vas, and Veszprém Counties. Nothing came of the discussions, and each county decided to deal with its own respective stretch. Such discord heightened flood risks at downstream places like Győr, at the confluence of two major rivers, because its local regulations alone could not prevent flooding.[24] When Franz I. called together Hungary's National Diet in 1825 after a thirteen-year lapse, the central body's ability to implement policies that would regulate the entire Middle Danube faced stiff resistance from local authorities (*alispánság*), who saw it as an effort to rescind their political autonomy. As local authorities lacked the financial resources to undertake engineering works themselves, more holistic approaches to flood control remained unresolved.

The early modern Habsburg Empire's disjointed approach to hydraulic engineering projects was in stark contrast to grand-scale regulation efforts envisioned elsewhere. In the Qing Empire, a well-established, albeit bloated technical administration was responsible for overseeing the dredging of rivers to keep them free for navigation and the building of embankments throughout the empire to hold back floodwaters. These efforts grew more and more expensive until, by one estimate, the water bureaucracy soaked up one-tenth of the government's budget by the nineteenth century.[25] Johann Gottfried Tulla's ambitious plans to regulate the entire Upper Rhine also envisioned an interconnected system of embankments, river transections, and regulations united under a single plan.[26] While coordinating and funding flood prevention measures was difficult for provincial authorities in the decentralized Habsburg Empire, a similar problem existed in America. Courts prevented the federal government from funding interstate infrastructure, so states, like their Habsburg provincial counterparts, had to rely on stock companies and tolls to raise funds for hydraulic projects.[27]

Major flooding in the early half of the century underscored uncoordinated and unfulfilled plans across the empire. In April 1830, when the Maros River in eastern Hungary overflowed, it flooded Torontal County. Water from the river reached the Danube and the capital of Bács-Bodrog County, Zombor, saw floodwater rip a two-hundred-fathom gap in the embankment, much to the dismay of numerous groups that had spent nearly forty years working on it. An article in the *Pressburger Zeitung* soberly stated: "It is undeniable that nature opposes human force and only allows itself to be guided by mortals. It is only through experienced hydro-technicians implementing a regularization of the Danube, the Tisza, the Vág, the Maros, and so on, not from the mere embanking of the rivers, that we in Hungary can expect the elimination of so frequent recurring and untold losses due to floods!"[28]

After the 1830s floods, the Hungarian government encouraged private initiatives to reduce flood dangers, allowing flood prevention companies (*árvízmentesítő társulatok*) to form. Nobles, entrepreneurs, and interested parties founded and funded such work, which drained and narrowed floodplains, built embankments and flood protections, and dug drainage canals.[29] This was an enormous undertaking in the Hungarian lands, where flood plains covered an estimated 14 percent of the territory.[30] Private initiatives, however, were inherently local or regional. In 1838, flooding wreaked havoc in Esztergom, Komárom, Győr, Pozsony, Buda, Pest, and Óbuda, destroying over ten thousand homes and killing three hundred people.

In the aftermath of the 1838 flood, the Hungarian government, with the support of Emperor Ferdinand (r. 1835–48), signaled that it would more actively pursue flood reduction measures. The National Diet passed Law IV in 1840 charging a newly formed "national council" (*oszágos válaszmány*) "to regulate the Danube and other rivers as well as the cities of Buda and Pest and their neighboring territory, for the sake of ensuring them against elemental vicissitudes."[31] Despite the organizational structure and support from the public, the 1840 law did not provide money to fund the council's work, and flooding in 1847 and 1849 highlighted its inactivity. Heading the Water Transport committee, István Széchenyi expressed satisfaction that the council had at least defined the Danube's regulation as part of the public's interest, a huge step as he saw it in reforming the funding mechanism for public works.[32]

Transnational Networks of Aid, Technology, and Science

Floods were a disruptive and destructive force, but in the nineteenth century they accelerated the forging of transnational networks across different regions of the empire. These social, political, and technical bonds revealed solidarity, cooperation, and coordination and united people together in a common desire to turn the tide on floods and flood danger. This transnational solidarity and cooperation were aided by education and growing interest in newspaper subscriptions, which enabled literate groups from around the empire to visualize and empathize with the plight of fellow citizens. The intricate reporting from numerous sources and within various provincial papers reveals the salience that natural disasters had for daily readers, which learned all about them in ever-greater detail. The stories of the summer flooding in June 1829, the March/Morva flooding in 1830 and again in 1838 rippled across the empire for weeks and months after each occurrence, saturating the news with stories of peril, destruction, heroism, and magnanimity. If newspapers were conduits through which groups reading about the events almost simultaneously across a shared geographic space created "imagined communities," then the Danube forged this community together by providing papers with tales of danger and solidarity.[33] Given the press censorship of the Metternich era after 1815 that discouraged overtly political themes, it is likely that natural occurrences garnered even more attention in the news.

Reading about the floods provided an opportunity for the literate public to imagine an empire-wide Danube space, with practical benefits in the aftermath of floods. While local communities undertook the arduous task of rebuilding when floodwaters receded, they also relied on relief pouring in from across the empire. Newspapers served to advertise and channel relief aid to afflicted parts of society. When the 1830 flood hit, the *Wiener Zeitung* published a large, front-page appeal asking for "noble-minded, well-intended people" to help with the monetary efforts to rebuild communities and get the poor classes affected back on their feet. The appeal reminded readers that *everyone* knew how these disasters affected people and expressed hope that people would feel compassion and help their fellow citizens again.[34]

At a time when smaller communities in particular had few, if any, resources to distribute for flood relief or to help rebuild, it was typical to elicit empire-wide aid.[35] People demonstrated great generosity with their donations in the aftermath of the 1830 floods: nearly 300,000 florins came from residents in Vienna, more than 75,000 from the various crownlands, and 3,000 from around Europe.[36] Lists of donations and stories of bravery appeared in the *Wiener Zeitung* the following day, as did an advertisement for a benefits concert to raise donations.[37] For weeks afterward, requests for donations spread throughout many local papers.[38] When a flood hit along the Middle Danube in 1838, residents in the town of Esztergom sent petitions to imperial institutions—the Palatine and the Austrian National Bank—for foodstuffs, relief materials, and funds to help rebuild. A local councilor recorded the town's gratitude for "the Viennese charity" (*a bécsi jótékony*), even as communities from neighboring areas likewise donated food and clothes.[39] In one of the most famous instances of flood relief efforts, Hungarian-born Franz Liszt broadly advertised his return to Vienna to give piano concerts to raise relief funds for flood victims of his native Hungary. Other institutions like museums, libraries, and associations emerged in the early nineteenth century as a critical source of funding and support for economic improvements and social causes that supported general well-being, like flood relief. Pieter M. Judson has argued that local institutions started such groups to improve community life, uniting many social classes and linguistic groups in the process.[40] Post-flood generosity demonstrated another facet of the empire's engaged and conscientious civil society.

Educational reforms and opportunities also underpinned the emergence of educational and natural associations focused on understanding the root causes of flooding. These associations worked with government agencies to record and publicize data on rainfall and water levels, in the hopes of discerning patterns between their meteorological and hydrological observations. The Societas Meteorologica Palatina/Ephemerides began publishing water levels on the Danube and Vltava as early as 1775 from its observation station at Buda and Prague. In 1784, the Imperial-Royal hydraulics administrator Jean-Baptiste

Brequin oversaw measurements of the Danube at Vienna, though his death the following year ended this experiment. Finally, in 1811, the *Wiener Zeitung* published a small excerpt in its paper discussing the barometric pressure and temperature at the Imperial-Royal Observatory in Vienna and the Danube's water level in the Viennese Canal for the first time. The *Wiener Zeitung*'s unassuming addition to its daily papers was soon joined by local papers in Buda (1817) and Pressburg/Pozsony (1819), while physical stations began recording the Danube's water levels at Linz (1821), Stein, Vienna-Nußdorf and Vienna-Kuchelau (1828), Melk (1831), Struden (1841), Tulln and Zwentendorf (1844), Fischamend and Hainburg (1846), and Mauthausen (1847).[41]

Thanks to the introduction and expansion of telegraphs in the 1840s, these far-flung observation stations were soon unified. Introduced in 1845, *Die Gegenwart* advertised a new "electromagnetic telegraph" that would run between Prague Castle and the Hofburg in Vienna.[42] Telegraph lines quickly sprouted up between Vienna, Brünn, Prague, and Pest, their construction hastened by floods in 1847 and 1849.[43] Indeed, as Michael Neundlinger has discussed, some of the first telegraph lines in the empire "were installed along the course of the Danube, connecting Upper Austria to Vienna. Telegrams were then used to send flood warnings from the upper stretches downstream to the capital city, where the *k.k. Telegrafen-Centrale* subsequently disseminated the warning to city dwellers."[44] The German physicist Karl August Steinheil (1801–70), chief of the newly established Telegraph Department in Vienna, oversaw the expansion of over one thousand miles of telegraph lines from 1849 to 1852 and these connected Vienna to all its provincial capitals and beyond.[45]

Franz Joseph's neoabsolutist rule (1850–59) supercharged efforts to forge an early warning system of these disparate water level observations by harmonizing technical administration of hydraulic affairs into one organ at Vienna. As part of this new warning system, the imperial government mandated that all provinces record scientific data, such as water speeds, river depths, and ice formation and send their observations to the central authorities. With the whole empire's data at hand, the imperial authorities created hydrological maps of the entire water system and was able to communicate potential flood dangers to any province forecasted to experience high water levels.[46] As data about water levels and precipitation grew more precise and integrated, they contributed mightily to the burgeoning field of climatology.

Understanding the diverse climatological relations between different provinces was one of several projects undertaken by naturalists and, later, scientists observing and recording data about physical and human geography in the Habsburg Empire. Deborah Coen writes about how these imperial-royal scholars chased after new insights in physics, biology, geology, and climatology in their efforts to discern a unifying principle behind the empire's diversity.[47] One such scholar to advance this work was Karl Kreil (1798–1862). Trained as an astronomer, Kreil was nevertheless exposed to many interdisciplinary ideas in his early career, rubbing shoulders with geologists, physicists, chemists, and other natural scientists. His pursuits led to his being named one of the inaugural members of the newly established Academy of Sciences in 1847, and from his new position he pushed the imperial authorities to establish a centralized meteorological institute. In 1851, Franz Joseph I approved this request, and thus the Imperial-Royal Central Institute for Meteorology and Earth Magnetism was formed. Named the institute's first director, Kreil set about assiduously assembling an empire-wide observation system that, like the River Engineering Office's plans for its river data, would unite the disparate observatories in the empire and make analysis of real-time data possible. According to Christa Hammerl, "One of Kreil's goals was to ensure that both the central meteorological station and the growing number of new meteorological stations throughout of the Austrian Empire were equipped with all the appropriate instruments. . . . The meteorological-station network, originally planned to have 100 stations across the empire, developed rapidly. In 1859, the network comprised 124 stations."[48]

The institute and its sprawling network began publishing daily weather maps in 1865, and by 1877 it was also forecasting weather for the following day.[49] The Central Institute's work was crucial for the accurate prediction of weather patterns affecting Danube water levels. The Hydrographical Central Bureau, established in 1893, depended on these observations for its own functioning. In Hungary, efforts to publish long-term and wide-scale observations led, in one instance, to a pastor from the Central Hungarian Plains using publicly available data about rainfall and water levels for several decades to try to mathematically calculate the exact relations between the two phenomena in Hungary.[50] Calculations became accurate enough that engineering departments published warning to residents when flood conditions remained in effect. For example, when a 1907 flood destroyed embankments near the confluence of the Danube and Tisza, flooding the town of Rudolfszgnád, the *Budapesti Hírlap* carried a warning from the National Hydraulic Engineering Department that high waters would continue to sweep down the Danube in the following days, based on reports and calculations from farther upstream.[51]

Communication infrastructure continued to advance with the invention of the telephone. While local groups had previously used telegraphs to contact imperial groups like the Imperial-Royal River Patrol (k.k. Stromaufsicht) to warn downstream communities during ice floes or floods and to mobilize rescue boats if necessary, in 1899, the imperial representative's office (*Statthalterei*) in Upper Austria raised the possibility of sharing costs to set up a more effective hydrographical service in Austria.[52] The Upper Austrian authorities set up a "hydrological services office" in Linz to monitor river conditions as part of the Flood Information Service. After October 1901, the building included a telephone station for warning different parts of the empire about dangerous water conditions that worked together with the Hydrographical Central Bureau.[53] By 1904, the Upper Austrian Landtag also opined that telephone stations should line each side of the Danube from Passau to Hainburg and cover the Danube's largest tributaries as well. Before this transition was complete, however, the Imperial-Royal Trade Ministry also received petitions requesting that citizens be able to send free telegraphs from any station along the Danube or its tributaries to warn of rising water levels.

Integrating technical breakthroughs and scientific discovery promised to unite the empire's inhabitants as they sought ways to overcome the surprise and devastation of flooding. This integration dovetailed with ad hoc but reliable networks of aid that emerged in the wake of tragedy to help people rebuild after times of flooding. Such transnational networks started as informal connections and grew into formalized institutions, which provided aid and warning to citizens of the empire. As we will see next, they also served as the foundations for more integrated prevention measures in the latter half of the nineteenth century as well as more coordinated responses from both governments and citizens.

Floods, Reform, and the Common Good

Just as floods necessitated technical and scientific developments to predict and protect against them, their threat and destruction also galvanized key reforms in the imperial government to unify and strengthen flood protection, response, and relief as a key issue for advancing the common good. Such work went hand in hand with centralizing efforts both in the empire and in many other parts of the world as key governmental responsibilities crystallized during this period. As Jürgen Osterhammel has pointed out, "It was not the early modern period but the nineteenth century that saw the transition from the traditional to the rational state. Inevitably this was bound up with the construction of bureaucracies and the expansion of state activity—a process observable almost everywhere in the world."[54]

The 1847 and 1849 floods were just some of the natural disasters that more generally took their toll on the empire's population in the mid-century, demonstrating the need for a stronger governmental role in protecting citizens. A potato blight in 1845 had caused an economic slump in 1846, and a cholera epidemic had also broken out in 1849.[55] Harvest failures in mid-decade also meant the cost of food constantly increased, as did the consumables tax on food items. Wood prices were so high that many people could not afford to heat their homes or cook their food. Wages, meanwhile, stagnated, and unemployment increased.[56] These events contributed to simmering working-class frustrations in cities and help explain, in part, peasant uprisings in Galicia in 1846. The result was a series of uprisings around the empire in 1848. While there is no denying the uprisings' national demands, many concerns simply signaled a general discontent with the authorities' perceived indifference to people's stagnating, and in some cases declining, socioeconomic fortunes.[57]

Faced with a deluge of crises, authorities sought ways to contain and turn back the social and natural threats. The possible elimination of floods' ubiquitous threat opened new avenues for the Habsburg regime to regain the populace's trust and loyalty, an issue that Hungarian revolutionaries protesting the imperial regime also took up. While petitions in 1848 did not specifically voice the need to protect people from floods several new programs initiated in Vienna and Buda sought to improve the strained relationship with the public and the empire's rivers. The revolutionary Hungarian cabinet and its new Ministry of Public Works and Transport committed to investing in hydraulic engineering projects to make Hungary's rivers safer. The imperial government at Vienna rechanneled frustrations by setting up "work programs," to send those without work to the Danube's banks, where they could earn money building up the river's embankments.⁵⁸ These carrots were not without their stick. Coming to power during the uprisings, Franz Joseph (r. 1848–1916) directed the imperial armies to subdue the internal unrest and by August 1849 with the help of Russian forces, the young emperor once again gained control of the empire's territories. In the revolution's aftermath, Franz Joseph consolidated political control and issued a raft of reforms to placate the discontent that had led to the 1848–49 uprisings. As part of these reforms, the imperial ministry shifted responsibility for river regulation and embankment construction from the provinces to the imperial capital. A newly established central office in Vienna within the Ministry for Trade, Industry, and Public Works unified and coordinated technical administration among the provinces.

While the imperial government took charge of coordinating regulation work, civil society played an active part in planning and implementing these changes. Public notables served with engineers as advisers on a Danube Regulation Commission drawing up plans to regulate the Danube at Vienna. The commission worked under the imperial trade minister Karl Ludwig von Bruck and consisted of ten engineers and six notables representing the interests of the city. In Buda and Pest, flood protection measures relied on cooperation between technical authorities and private initiatives. While engineers in the new Imperial-Royal Engineering Directorate in Buda began the critical task of surveying the islands and low-lying land adjacent to the river to determine the risk of flooding, the directorate collaborated with private groups like the Danube Steam Navigation Company (DDSG) and the Embankment and Canal Construction Company to build quays and embankments along this stretch of the Danube.⁵⁹ Other private companies drained and regulated stretches to further reduce flood risks.

The government's partnerships with companies harnessed the dynamism of commercial and technical groups. Such a shift reflected Franz Joseph's emulation of Josephinist reforms, which promoted centralized bureaucratic rule and technocratic expertise at the expense of the aristocracy-dominated provincial diets. The pitfall of relying on "experts" was that technical disagreements could, and did, hamper progress. In 1858, an engineer at the Engineering Directorate in Buda expounded these challenges in a series of articles for the *Pesti Napló*'s considerable readership. Explaining the specific seasonal and hydrological conditions that caused recurring flooding in Buda and Pest, the engineer József Péch then outlined steps that engineers hoped to undertake to mitigate these factors. His articles described the difficulty of uniting around any singular regulation plan, given the contradictory technical opinions on how best to satisfy interests in navigation, flood protection, and land reclamation.[60] Such sentiments could likewise have emanated from representatives of the Danube Regulation Commission in Vienna the following decade. In 1864, the emperor formed a commission to devise a better plan for protecting Vienna. The commission united technical experts with representatives from the new Imperial Diet, the Lower Austrian provincial diet, the Chamber of Commerce, the DDSG, and the Northern Railway Company. Disagreements about the most effective plan to reduce flood dangers eventually prompted Chief Engineer Pasetti to implement his own plan: leave the river in its existing bed.

While authorities focused much of their attention on regulating the Danube at Vienna and Pest-Buda (as of 1873 Budapest), embankment works also spread along other stretches of the Danube. Requisitioning large amounts of stone from quarries between Passau and Theben/Dévény, the imperial authorities provided local communities with construction material in order to speed up the process of approving and building embankments.[61] In 1858, Carl Freiherr von Czoernig praised imperial plans and expenditures that supported the unification of levee projects across the empire stating that "the hitherto successes [in regulating the river] speak to the correctness of these plans."[62] From 1850 to 1855, levee construction reached approximately 150 miles and by 1861 it was closer to 300 miles on just the Upper and Lower Austrian stretches of the Danube.[63] Embankment construction and river improvements grew beyond the Danube and included works on the Weichsel (Vistula), the Dunajec, the San and Dneister in Galicia, and the regulation of the Vltava (Moldau) and Elbe in Bohemia. In 1850, the emperor also decreed an annual sum of 100,000 florins to cover costs for the Tisza's regulation to reduce flooding along the Danube's largest tributary. In 1856, he extended this mandate to include the protection of floodplains along the Tisza's major tributaries as well. Despite these efforts, a series of floods in 1850, 1852, 1858, and 1860 in the Middle Danube and Upper Rába featured widespread despair over the destruction that continued to afflict riparian communities. A book released in the aftermath of the 1862 flood lamented "the swelling of

Overcoming Danubian Dangers

the Ottakring Brook in the night of January 30, 1862, and the torrent that struck part of the district Neubau [were] simply the prelude to subsequent disasters affecting not only Vienna and its surroundings but a considerable stretch of land that encompassed millions of the empire's inhabitants."[64]

This destruction had a lasting political legacy in the coming decades. After the 1867 Ausgleich, new imperial and national legislatures quickly approved plans for raising funds and undertaking additional regulation works at Vienna and Pest-Buda. In February 1871, Baron Gyula Nyáry also implored his colleagues in the House of Magnates to consider flood prevention and flood relief measures for the counties: "Perhaps when we consider Budapest, water danger has largely subsided, but while communities along the Danube are not yet saved from the threat of flooding—allegedly Tolna environs are already under water—I venture to ask the members of this high office: shall we not find out what precautionary measures and rescue and relief arrangements should first be taken care of?"[65] In response, the Ministry of Public Works and Transport founded the River Engineering Office, which had thirteen offices on the Danube and "supervised waterways, provided hydrographic duties, checked regulated river stretches, and directed flood protection measures."[66] The National Diet also passed legislation (Law XXXIX) in 1871 to regulate the formation of flood protection companies, which had proliferated after 1867. These companies undertook regulation and drainage projects on all major waterways in Hungary. In addition, approximately six to ten embankment construction companies (*gátépitési társulatok*) bolstered efforts to keep floods at bay.

In the capitals, regulation works were completed from 1870 to 1875, and freezing conditions soon tested the cities' new, protective measures. The winter from 1875 to 1876 was so cold that ice had started flowing down the Danube near Budapest in December, piling up in unregulated stretches south of the city, blocking the river's flow, and threatening to raise water levels over the city's embankments. Reports from the county authorities and technical-hydrological supervisors in different cities along the river also began pouring into the capital.[67] The Vienna-based newspaper *Die Presse* tried to reassure readers that new measures to reduce flood damages and drain floodwaters quickly would protect citizens, though it also warned them to take precautions should the city unexpectedly flood.[68] The papers posted the ice's progress every day. One article quoted the engineers who had designed the flood protective measures as cautioning that a "complete guarantee of safety [was] beyond the ability of man."[69] By the end of January, defenses seemed to be holding, and the papers tentatively celebrated the passing of danger. When temperatures plunged unexpectedly, the Danube froze solid for weeks. Warm weather in mid-February thawed the ice, causing chunks to ominously pile up and block the river. Melted snow likewise swelled the river, and water levels rose precipitously, flooding tens of thousands of square kilometers across the empire.

When the floodwaters eventually subsided, the blame game started. Hungarian newspapers negatively compared the damages to the 1838 flood but argued that while the two floods caused similar levels of damage, the flood in 1838 lasted only a few days, whereas floodwater in 1876 lingered for weeks. The chief Danube engineer at Vienna, Gustav Ritter von Wex, acknowledged that such natural disasters remained a threat to all riparian communities, reminding his audience that the recent thaw had also affected the Rhine, Elbe, and Seine, where "many cities, and even Paris, suffered from higher and more destructive floods than have occurred since the last century."[70]

However, Wex and other engineers were soon forced to realize that their interventions had also played a part in magnifying these disasters. In 1880, the engineer Károly Hieronymi wrote that many had debated the soundness of regulation plans at Budapest after the 1876 flood, though new regulatory works in 1881 continued to follow the same precept of one river, one bed.[71] In Vienna, engineers had actually expressed doubt *during* the 1870–75 construction that a single new bed could handle the projected volume of larger floods. Following the flood, Wex published his calculations concluding that embankments along the Viennese Danube, rather than protecting the city's inhabitants, had ironically raised water levels and made flooding more likely.[72] Ideas to create a relief channel parallel to the main bed of the Danube remained unfulfilled at Vienna until the 1970s.[73]

The capitals were not alone in their tragedies; a flood of petitions and natural disasters prompted legislatures to designate funds for regulating rivers and draining floodplains along several Danube tributaries as well. Factory owners in communities like Schwertberg, Mistlberg, and Altaist as well as residents from along the entire Aist River wanted to see what measures the state could undertake to protect it from the "danger of new, disastrous floods."[74] In 1881, the Upper Austrian Trade and Industry Association castigated the lack of embankment protection along the Inn River, the empire's boundary with Bavaria in the German Empire. The association claimed that Bavaria's decades-long embankment works "routinely deflected flood waters into the empire" and "destroyed large tracts of arable land."[75] In 1882, heavy rains and mountain torrents in Carinthia and Tirol led to avalanches and flooding and some of the worst death tolls in the Habsburg Empire from natural catastrophes.[76] This last occurrence finally prompted imperial legislators, in conjunction with provincial diets, to pass several laws guaranteeing joint funding for the regulation of prominent rivers and mountain streams. A flood on the Rába River in 1883 likewise precipitated the passage of Law XV in Budapest wherein the National Diet provided seven million florins for the river's regulation and embankment. Securing the support of central governments in Vienna and Budapest was critical for funding large, expensive provincial projects. When local communities were unable to secure their residents from flooding on the Gail River, petitioners acknowledged that "the [provincial] government now considers that support from the [imperial] state alone will lead to the desired goal."[77]

Much of this new legislation reflected a shift in rationale for river regulation projects, which moved past strong navigational incentives and instead focused on attenuating flood dangers. Flooding on the Danube, Maros, Tisza, and Rába Rivers in 1879, 1881, and 1883 led executive and legislative bodies in Hungary to fund plans to jointly manage flood protections and secure floodplains. The National Diet's 1885 law regulating flood protection companies also placed a floodplain development tax on landowners to provide funds for constructing and maintaining embankments and drainage canals. By 1890, in its push to reclaim floodplains, build embankments, and regulate rivers, the Hungarian parliament had invested over 100 million florins. Meanwhile, flood protection companies had excavated 100 million cubic meters of land (an amount of soil that would fill one hundred Empire State Buildings) to straighten riverbeds.[78]

These political solutions succeeded because active citizens and local communities took part in advocating their visions for the regulated river. When the Lower Austrian provincial diet passed legislation to regulate its stretch of the Danube, the city council at Melk protested the official plans, believing they would leave the town open to additional flooding. Council members petitioned the Danube Association to intervene with the imperial authorities. When heavy flooding in 1897 and 1899 sped along plans to complete the Danube's regulation, the municipal council again presented its plans tailored to "best protect the city's property." The undertaking authority, the Danube Regulation Commission, eventually respected the wishes of Melk's residents. The council also managed to secure funds from the district offices to raise an embankment along the local alluvial plains.

Businesses and civil associations teamed up to help the government plan and oversee hydraulic projects. The Danube Steam Navigation Companies coordinated with the Danube Regulation Commission and several imperial ministries in the wake of flooding disasters to monitor reconstruction work of embankments up and down the Danube. In November 1899, the Danube Regulation Commission decided to sponsor a River Viewing Trip (*Stromschaufahrt*) to visit sites of flood damage and direct measures to prevent future flooding.[79] Participants set off in May and concluded that it was necessary to embank the Danube from Melk to the Hungarian border.[80] The Trade and Interior Ministries sponsored an additional trip in 1902 while low water levels in spring 1906 prompted yet another trip.[81] That fall, representatives from the DDSG, the Donauverein, the Imperial-Royal Trade, Interior, and War Ministries, and Bavarian delegates spent another few days examining the entire length of the Danube from Passau to the Hungarian border.[82] All told, these trips continued every few years, lasting into the time of World War I with a trip along the Moldau taking place in 1915.

Despite painfully slow progress along some stretches, in the last few decades of the empire, work continued with the goal in mind of keeping residents safe. In the empire's western half, provincial authorities bolstered the work done by imperial ministries. The Upper Austrian executive and legislative branches listed the Danube tributaries (and tributaries' tributaries) that they expected the imperial authorities to regulate, estimating a cost of nearly 4 million crowns (2 million florins before an 1892 currency reform pegged the empire to the gold standard).[83] To cover half these expenses, the Upper Austrian provincial authorities committed to providing 1 million crowns and expected local governments to provide another million. The Upper Austrian diet then petitioned the imperial authorities to mobilize the Meliorations-Fond to cover the other half. After the 1899 flood, the Danube Regulation Commission in Lower Austria reported a 50 percent rise in funds covering new flood embankments and repairs for flood-stricken communities.[84] Upper and Lower Austrian were alpine provinces and representatives from each recognized that to avoid future flooding, they needed to regulate rivers and construct "waterways" in the mountains to prevent rain and glacial melt waters from becoming uncontrollable mountain torrents.

The Hungarian government, for its part, increased embankment and protection measures, including reclaiming alluvial floodplains to an impressive degree. Flood protection and river regulation companies invested 345 million crowns to improve protection, and from 1867 to 1905, the National Diet invested 230 million crowns on regulation work and another 30 million maintaining its hydraulic engineering works along the Danube, Tisza, Szamos, Bodrog, Kőrös, Temes, and Bega Rivers. In 1908, a representative from the House of Representatives put these accomplishments into perspective. Iván Reök spoke to the chamber, pointing out that the Po River's regulation and land reclamation had long served as an example to emulate in Hungary, as had the Netherlands' land reclamation works, and France's Loire Valley regulation. Reök proudly stated that while those three projects had protected a total of 2.1 million hectares of land, the investments in Hungary along the Danube and Tisza Valleys had embanked and protected an impressive 3.6 million hectares (about the size of Connecticut or Taiwan).[85] By 1914, the Danube River in Hungary alone had a total of 38 flood protection companies, which had reclaimed over 1 million hectares of land, and dug almost 2,500 miles of canals to aid drainage and irrigation.[86]

Reclamation projects had mixed economic results. Drainage projects increased the amount of arable land in Hungary, which helped large and middle-sized landowners. Former serfs also cultivated newly reclaimed plots. However, in some places, such as the Bodrogköz (located in the Upper Tisza region between the Tisza and Boldrog Rivers), converting land from marsh and pasture to arable land in the 1890s came too late to benefit from the earlier era of high grain prices. Likewise, the retreat of animal husbandry in the fens and marsh pastures reduced an important source of fertilizer. At the same time, traditional cultivation methods continued to be prevalent, reducing crop yields and the value of newly won land. According to Balázs Borsos, "Social and economic factors that prompted the natural and agricultural changes at the same time hindered the complete utilization of the new natural environment."[87]

From an ecological perspective, drainage and regulation works were decidedly negative for many peasants who relied on rivers' water, flora, and fauna as the basis for husbandry and agricultural pursuits. For centuries, peasants had possessed the legal right to use the fanning floodplain wetlands as commons where their livestock could avail themselves of roots, shoots, and even small fauna. Galley forests in certain stretches provided acorns for masting pigs. While few staple crops could be cultivated in the regions that frequently flooded, peasants were well accustomed to foraging edible vegetations and tending to orchards in higher, drier portions of the floodplains.[88] Ethnographers have documented that many peasants, particularly those who did not own any land (around 40 percent of the peasantry in the nineteenth century), relied on these riverscapes.[89] When wars against the French once again drove up the price of grain in the early nineteenth century, aristocrats and large landowners invested in drainage and regulation projects to increase more valuable grain-producing land. This process shifted production away from cattle raising and toward grain cultivation, squeezing peasants' access to commons.[90] Nobility in eastern Félvidék (today's Slovakia) confiscated large tracts of land during this process—in some places nearly 60 percent.[91] After an 1830 crop failure and 1831 cholera outbreak inflamed local peasants, one of the largest uprisings broke out along the Boldrog River region and spread, with forty thousand peasants demanding the division of noble estates, freedom from feudal obligations, and self-governance.[92]

Besides river husbandry, the unregulated Danube with its slow, shallow meanders, side arms, semiaquatic wetlands, and oxbow lakes also provided a rich ecosystem for fishing communities. Broadly speaking, the rights to fish the river were owned by adjacent landowners. In the medieval and early modern period, they leased these rights out to fishing communities and guilds.[93] Fish was a necessary staple during the Lenten months, and fishers relied on large markets in growing cities up and down the river to sell their catches. To catch fish, communities developed techniques that used the natural dynamics of unregulated and flood-prone rivers. Bertalan Andrásfalvy has studied alluvial plains and bank dwellers, arguing that villagers and peasants excavated and maintained small side channels in the Danube called *fok*, where fish swam and spawned. Using stick weirs and nets, fishers caught larger fish while smaller fry escaped and were able to populate the river anew. Floodwaters replenished the *fok*.[94] Local fishers credited this practice as the basis for Hungarian rivers' bountiful supply of fish.[95]

But because engineers were quick to blame unregulated rivers for flooding more readily, they disregarded these local practices, cutting off these side channels in their quest to engineer the river to make it safer and more profitable. In the same way, when flood reduction companies (*árvizmentesítő tarsulatok*) were formed in the 1830s, they bought land adjacent to the river from landowners to begin the process of regulating segments of the Danube, building embankments, and draining the adjacent wetlands. For fishing communities, the ecological disruptions that accompanied these works were matched by economic dislocations: drainage and embankment companies typically bought only the land adjacent to rivers but did not purchase the "water usage rights" attached to them. This muddled the legality regarding who could fish from the newly reclaimed properties up and down the river, galvanizing fishing associations and fishing congresses to try to address this issue from the passage of the ambiguous 1888 fishing law onward.[96]

Despite the adverse effects for certain segments of society, drainage and flood reduction companies mobilized the winning argument that their work benefited the economic and material well-being of the public. It was a convincing perspective to many who suffered fear, inconvenience, or financial loss because of the river. In a speech before the Pressburg Association for Natural Sciences, Dr. Ruprecht exclaimed of the Pressburgh Csillaköz Island drainage, "It is not an inconsiderable value gained when looking at the trees planted in the previously barren landscape . . . and the more easily navigable pathway through which boats can now sail from village to village in what had previously been impassable swamps . . . [and] the incalculable value of more sanitary conditions by ridding this region of the fever epidemics that had struck those working in the fields here. . . . It is thus a joy for any friend of progress that this patriotic work continues."[97] Nature needed to be subdued, not accommodated.

Imperial Behavior, Active Citizens, and Institutional Relief

Engineers transformed the empire's physical environments in their quest to eliminate the multifaceted causes of floods. Communities and officials, however, recognized that protective measures alone would not solve the problems that floods caused. Instead, clubs, committees, and concerned citizens banded together and established new local and institutional approaches to preparing for and responding to floods. New protocols, new behaviors, and new agencies all represented a deepening of the government's obligations toward its citizens and a broadening of citizens' responsibilities to guarantee the "public good" during times of natural disasters.

Starting with flood protection and responses, imperial and local authorities began to cooperate and coordinate their efforts to prepare for floods and to respond in an organized fashion during rescue operations. Franz Joseph's imperial representatives (*Statthalter*) in the crownlands oversaw embankment construction and coordinated with local governments to develop government protocols for responding to flood crises. These protocols included delineating distinct imperial and municipal duties during times of floods. In 1851, the *Statthalter* for Lower Austria issued an edict clarifying the responsibilities of the Viennese mayoral office, the Lower Austrian Engineering Directorate, the Viennese municipal authorities, and half a dozen district leaders in Lower Austria in the event of a flood. The *Statthalter*'s office in Hungary wrote to the Pest mayor with instructions for the authorities in both Pest and Buda. Outlining its own duties for protecting citizens, the *Statthalter* then designated the police directorate (*cs.k. rendőrigazgatóság*) as the authority for preserving "public calm and order" and charged the Buda and Pest councils with maintaining communication channels to coordinate search and rescue efforts and distribute aid. The engineering directorate would be responsible for undertaking technical repairs during the flood.[98]

Chains of command gained additional support from new associations and governmental committees which jointly monitored water levels, trained rescue volunteers, and stockpiled reserve supplies for flood aid. In 1860, the office of the imperial representative (*Statthalterei*) in Hungary created a commission, which included representatives from the mayor's office and engineering directorate, to oversee and coordinate subcommissions in each district. These subcommissions were responsible for ensuring that the proper number of rescue boats were on hand, designating locations for food distribution, and determining which buildings might not survive sustained flooding. After the 1862 flood hit Vienna, the communal council followed suit, setting up the Advisory Commission for Protective Measures Against Flood Dangers. This commission operated for decades, ordering rescue boats, overseeing levee construction, scheduling water level patrols, ordering food aid reserves, and designating buildings to house flood victims who had lost their homes.

Complementing government oversight and management, citizens participated in preventative measures. Businesses and individuals removed rafts and ships from the Vienna Canal whenever Danube navigation ended for the season to avoid obstacles that could cause flooding. The city magistrate commandeered left-behind vessels to deter noncompliance.[99] In Buda, authorities sounded cannons to signal danger. They also served as a call to "arms" for those involved in rescue efforts. Crews from fishing and transport ships were expected to make themselves and their crafts available during rescue operations.

Volunteer groups proliferated around the empire to train and assist in flood rescue operations. In 1868, Vienna's residents organized 163 rescue ships, and 380 men volunteered for the new water brigade (*Wasserwehr*).[100] When ice floes on the Danube threatened to flood Budapest in winter 1875, the *Vásárnapi Újság* reported that a Voluntary Rescue Committee of 110 members had taken part in initial relief and safety efforts. Viennese authorities brought together representatives of municipalities along the river to form "safety defense forces" (*Sicherheitswache*).[101] By February when flooding threatened again, a new Central Inspectorate for Safety Defense assembled hundreds of volunteers and boats to serve as rescue crews and the *Statthalterei* put troops from provincial cities like Linz and Pressburg/Pozsony at Vienna's disposal if they were needed.[102] When floodwaters did spill over the embankments at Vienna, residents formed chains to clear water while cannon fire signaled the highest level of danger.[103]

Likewise, smaller communities developed a series of official networks to activate in the event of a flood. In Tirol, flooding and mountain torrents in 1883 prompted a more systematic approach to training and preparing disaster rescue corps. In 1886, the Lower Austrian fire brigade bulletin cited Tirol's example and articulated a similar need, "lacking equipment, several places have resisted in vain the floods of recent years, which has turned our thoughts toward creating an organization similar to the fire brigade which will allow us to fight against such natural disasters with greater success."[104] In 1890, citizens in Melk founded a Flood Committee to implement ideas for reducing flood dangers. By 1899, the committee had established specific protocols to protect residents, including plans for safely moving the swimming facilities out of the river and notifying residents in low-lying areas of impending flooding. After the 1899 floods, Linz and Wels also established volunteer water brigades out of local fire brigades. These brigades brought together diverse members of the public including city councilors, members of rowing clubs and gymnastics associations, and civil engineers.[105] By 1913, Linz had almost fifty men serving in the water brigade. All around the empire, volunteers helped establish these rescue associations. Local initiatives received support from provincial and imperial authorities, who subsidized equipment for these groups in the same manner that it did for fire brigades.[106]

Massive flooding in the last decades of the century tested these new behaviors and protocols in profound ways. When the Vltava flooded at Prague and a few days later the Danube did so from Passau to Pressburg/ Pozsony, tens of thousands of people lost their homes.[107] In Vienna alone over 80,000 people had to stay in military barracks, hotels, workshops, and factories.[108] During the flooding, Franz Joseph went to the worst-hit districts to oversee relief efforts and to ensure orderly recordkeeping for compensating the victims.[109] He also coordinated with the military to lead search and rescue missions. In 1876, disastrous flooding covered nearly 4,000 square miles of land, amounting to one-quarter of the Carpathian Basin's floodplains.[110] Water lingered in towns and the countryside for over a month. Over 15,000 people in Hungary lost their homes and eventually headed to Budapest for aid. Floods in 1883 likewise destroyed hundreds of houses and left over 10,000 people without shelter. Two summer floods in 1897 and 1899 were devastating.[111]

Tragedy and loss revealed the persistence of transnational generosity and solidarity and strengthened the institutional ties that bound together the empire's residents. A few days after the 1862 floods, the Imperial Diet's president announced that delegates who wished to raise funds for the "kingdoms and provinces" affected by flooding could sell subscriptions and receive donations, which the State Ministry would distribute throughout the empire.[112] Following the 1876 flooding throughout the empire, the interior minister sent 8,000 florins for relief, and societal donations amounted to another 4,000 florins for the Kisföld alone. Franz Joseph I, Elisabeth, and other members of the dynastic family donated nearly 80,000 florins. The Rothschilds gave 5,000 florins, notables in Hungary collected a respectable amount, and Franz Joseph and Elisabeth donated an additional 40,000 florins once assessments of the damage became clearer. In the Austrian half of the empire, donations were followed up with imperial loans—willed by the emperor and legislated by the House of Representatives—to help communities rebuild. After the 1883 flooding in Hungary, rescue operations demonstrated that more sophisticated practices had also developed outside the capitals. Relief measures, however, remained ad hoc; committees formed to organize aid, the interior minister prompted the National Diet to pass relief legislation, private groups donated clothes, food, and money, and Franz Joseph I donated funds.

Finally, by the 1880s, representative bodies and government agencies began establishing formal protocols for collecting and distributing relief and reconstruction funds to flood-stricken communities. In 1882, 1897, 1899, and 1900, Franz Joseph I empowered the Imperial Diet to approve loans to communities that needed reconstruction funds. In 1897, this action amounted to an interest-free loan of 1 million crowns as well as an additional 300,000 crowns to rebuild embankments. After the flooding in 1899, the Imperial Diet approved 3 million crowns in reconstruction loans and donations to provinces across the empire. In December 1900, Franz Joseph decreed that Bohemia, Silesia, Upper and Lower Austria, and Vorarlberg receive 2 million crowns in flood relief. Provincial authorities also released funds, recognizing that supporting people in times of need was beneficial for both "economic and financial reasons."[113] The government's immediate and coordinated action was necessary; after Danube flooding in 1896, damages exceeded 250,000 crowns in Upper Austria, for which the Landtag was able to muster only 19,000 crowns for relief. After 1897, the provincial diet passed a province-wide emergency provision to help those affected, which included the lessening of military duties and taxes, the preferential treatment of businesses affected by the flooding, and a discussion about offering emergency credit to those in need of financial relief.

While the government mobilized funds for aid and reconstruction, civil associations assisted with efforts to distribute flood relief. In 1886, the Upper Austrian provincial diet set up councils and district communes to educate farmers and advocate for them at the political level. By 1895, there were 58 such collectives with over 11,000 members. When the 1897 and 1899 floods occurred, the Upper Austrian *Statthalterei* donated over 45,000 crowns to support farmers facing emergency conditions, and the Agriculture Ministry donated another 10,000 crowns. The Imperial-Royal Agriculture Ministry charged members of these councils with distributing funds to alleviate losses to farmers whose harvests had been affected by floodwaters.[114]

Throughout the nineteenth century and especially in the second half of the century, the persistent threat of floods promoted the establishment of new governing institutions that sought to protect the material well-being of the empire's inhabitants and help them rebuild in the aftermath of flooding destruction. Such governmental responses mirrored global developments in bureaucracy-building that enabled states to more effectively react to crises and mobilize help on behalf of their citizens. Threats due to natural disasters also animated new behaviors among local populations, associations, and businesses, which prepared more institutionalized responses to ward off flood dangers and distribute post-flood relief. Collectively, this work guaranteed strong partnerships between governments and local populaces and relied on the ongoing participation of citizens and volunteers who were willing to work to protect the well-being of riparian communities throughout the empire.

Flood Prevention, the Common Good, and a Return to Nature

When the Bretterwandbach once again flooded the Tirolean village of Windisch-Matrei in 1895, the requisite authorities were dispatched to survey the damages. Dozens of houses were washed away and over one hundred acres of its best agricultural land was destroyed. Unlike in 1849, however, there were no self-proclaimed cameralists criticizing governmental inaction or lecturing on the responsibilities of the citizenry. Instead, a new imperial law of 1884 made funds available through the Agricultural Ministry to help communities recover from mountain torrents and avalanches. The Tirolean provincial diet in Innsbruck contacted the imperial authorities on the village's behalf to release funds to help it rebuild. Much had changed in just a few short decades.

Nevertheless, a century and a half after major regulation works, the Danube today still represents a risk for those living along it. Ever-higher concrete embankments have cut the river off from its natural floodplains, which had previously served to absorb and disperse floodwaters. The Danube's channelization has increased its speed, raising the height of floods during times of greater water volume. Citizens settled more and more within historical floodplains, placing their faith in technical assurances and physical structures that promise to hold back any deluges. These "former" floodplains are still the first places to flood, and the Danube's historic floods in 1954 and 2013 demonstrate that engineering works did not eliminate the river's potential for danger.[115]

Discussions around mitigating flooding have taken a surprising turn—rewilding of the Danube. The website of the Danube Floodplains National Park (Nationalpark Donau Auen) downstream from Vienna asks readers, "Can floodplain landscapes contribute to flood protection?" It goes on to answer "yes," explaining that these watery landscapes help absorb and slowly release floodwaters, reducing the worst effects of flooding. The website concludes that "rivers need space to expand in the event of floods. It is therefore necessary to complement construction and improvement of levees and other protective measures with the preservation or reconstruction of unpopulated floodplains and retention areas along the entire Danube and other rivers."[116] Such conclusions run counter to the prevailing nineteenth-century wisdom that advocated for the channelization of the river.

Like earlier efforts, this new approach to flood prevention depends heavily on transnational cooperation. The European Union has promoted the rewilding of the Lower Danube to increase Europe's resiliency in the face of climate change. According to the website, "In the Lower Danube Green Corridor Agreement the governments of Bulgaria, Romania, Moldova and Ukraine agreed to restore 224,000 hectares of floodplain.... Decommissioning under-performing flood protection dikes and restoring floodplains is contributing to safer and more effective floodwater retention.... During the 2013 flood in the Danube, along the Lower Danube there was no flooding, although water was above the average level."[117] This would not have been possible without these states working together.

In this regard, the Habsburg Empire was ahead of its time, attempting to solve large, complex problems, like flooding, by relying on cooperation and collaboration across its vast empire. When political innovations, economic means, and technical knowledge in the nineteenth century promised to radically transform the Danube and its tributaries into a safer space to inhabit, governments, companies, and individuals leaped at the opportunity to advance these plans. Executive and legislative bodies from the local to imperial-royal level invested in regulation, drainage, and embankment efforts and worked alongside companies to promote the physical-material well-being of the public. As engineers slowly recognized that engineering works alone could not keep floods at bay, new institutions and regulations sought to codify behaviors and government mechanisms to both mitigate flooding and more quickly rebuild in its wake. Avenues of civic engagement, such as petitions, volunteering, donations, and representative bodies, allowed citizens to play an active role in this process, shaping and supporting plans to keep themselves and their communities safe.

Like so many river engineering works from the Rhine to the Mississippi, flood preventative measures were not without their ecological and social consequences.[118] Efforts to drain floodplains and reduce standing water, seen as flood multipliers, consequently lowered water table levels, magnified the effect of droughts, and increased the expansion of alkali soils in places such as the Hortóbagy in the Great Hungarian Plains.[119] Farmers alternatively complained about floods bringing too much water and drainage causing too little. Fishing communities and peasants relied on the unregulated Danube to provide plentiful fish and grazing commons, and they grappled with the loss of these traditional riverscapes once the Danube was straightened and its floodplains cut off and drained. A contemporary dissertation on Hungary's agriculture claimed that peasants' dispossession from the land in this period "fueled the Agro-Socialist movement" and "drove emigration."[120]

Despite drawbacks, proponents depicted this work as advancing "general well-being," while governments recognized that positive outcomes underpinned public approbation of their governance. By overseeing projects to prevent flooding and guarantee funds to help rebuild after natural disasters, authorities in the Habsburg Empire sought to secure what governments in this era of mass politics and democracy increasingly pursued from their people: trust in state institutions and loyalty to them. While money and government support could not prevent the occurrence of all floods, they did strengthen the imperial government's ability to counter the worst effects of flooding and slowly turn the tide against the vicissitudes of nature affecting their citizens.

Chapter 5

ACT LOCALLY, THINK IMPERIALLY

The visitor who sees Budapest for the first time is surprised by the beautiful layout of the double city. The Buda Royal Palace is prominent on its proud height, Gellért Hill looks out over the unbounded plains of the Alföld, the violent Danube flows down between two rows of palaces and under three permanent bridges, the middle of which is the Chain Bridge, a masterpiece of the monumental bridge construction, and in the river's center calmly sits the Margaret Island, while a swarm of steamboats rush about. Smoking chimneys advertise the capital's developed factory industry, and swarming workforces on the quays affirm the city's burgeoning trade.

—**Mór Jókai**

The prominent Hungarian writer and statesman Mór Jókai penned this vignette of Budapest to describe how, by the end of the nineteenth century, the city had become "the heart of Hungary." This paragraph reveals a city transformed by new quays, factories, and bridges and enlivened by its modern steamboat travel and "swarming" quay workforce. The initial focus is not on anything specifically *national* about the city's character. The Royal Palace, renovated and ceremoniously reopened by Franz Joseph in 1893, alluded to the Habsburgs' traditional sovereignty over Hungary's multiethnic territory. The description of factory chimneys, bridges, and quay workforces could have described most industrialized river cities around the world from London to Pittsburgh.[1] The "swarm" of steamboats described "rush[ing] about" were certainly a diverse mix of boats. Local ferries crisscrossed the Danube bringing workers and visitors to various landing places in the city. Steamboats from several regional companies picked up commuters and goods traveling from the capital to nearby towns. A few stray boats may have been arriving from Germany, Serbia, or Romania. But the most visible boats would be from the Danube Steam Navigation Company's (DDSG) fleet of hundreds of ships, with its passenger lines connecting Budapest to cities across the empire and throughout Central and Southeastern Europe. At the center of everything, the Danube River served as the scene's nexus between the local, transnational, and international elements.[2]

Many cities on the Danube included a cosmopolitan mixture of cultures, languages, and people, who had settled in larger commercial cities over the centuries, and nineteenth-century engineering projects magnified this heterogeneity.[3] Regulation, embankment, and drainage projects reclaimed and secured large swaths of land for urban growth near the Danube. Many riparian cities built new industrial zones, such as mills, factories, and processing plants, which offered opportunities for a steady stream of internal migrants arriving from around the empire.[4] While in 1857 Vienna had approximately 100,000 internal migrants, by 1910 that number had increased to 470,000.[5] Southern Bohemians were also drawn to nearby industrial towns in Austria's heartland, such as Linz and Steyr, located on the Danube and one of its tributaries, respectively.[6] One nineteenth-century source described migration in Hungary "in the other parts of the country the municipalities owe their growth almost entirely to immigration. As central points of industry, commerce and communication they absorb such of the population of the neighbourhood as are seeking a livelihood so that the greater part of their inhabitants are born elsewhere. This was the case according to the census of 1891 of not less than 50–60%

of the civil population of Fiume, Győr, Kassa, Koloszvár, Nagyvárad, Pécs, Pozsony, and Temesvár."[7] Nearly all these cities were located on the Danube, one of its tributaries, or a tributary's tributary. Migration to Budapest drove population growth in the mid-nineteenth century as well. In the late 1860s, every third person in the city was a newcomer.[8] Religious diversity increased as Jewish citizens gained unrestricted freedom to reside in larger cities after the 1848 revolutions. By 1910, Vienna had 175,000 Jews, and one-quarter of Budapest's population was Jewish.[9]

As cities grew and transformed physically to contain this influx of people, their relationship with the Danube adapted to accommodate intersecting imperial, transnational, and local interests.[10] As Nathaniel D. Wood has argued about cities in the Habsburg and Ottoman Empires, "While it is indisputable that many ... were becoming increasingly national during this period, they were also becoming 'modern' and modernity was, and remains, something much larger than nationality."[11] As cities large and small transformed in these decades, industrial, commercial, even recreational activities along the Danube remained rooted in pragmatic concerns about local river conditions, concerns more apt to be addressed by imperial or transnational cooperation than nationalist mobilization.

Regulation works enhanced people's connection to the Danube by providing municipal inhabitants with improved access to beneficial commercial, sanitary, and recreational activities. These projects were replicated throughout the empire, and governments from the local to the imperial level pursued them to support the well-being of the empire's inhabitants. Engineering projects along the Danube affected residents' relationships with governmental, associational, transnational, and commercial groups in the empire. Local populations were often at the mercy of the Danube's larger hydrological forces extending beyond their municipal boundaries. As a result of this dynamic, people used civic and imperial avenues via petitions, appeals to representative bodies and private enterprises, and through civil associations, to protect and advance their local interests as they saw them embedded in the Danube.

Industrial-Commercial Zones

In 1896, a general work about the Danube, *The Danube as Cultural Path, Navigation Avenue, and Travel Route*, described the regulation works undertaken at Vienna earlier in the century. According to the author, engineering the Danube was meant to "meet the present and future needs of trade, transportation, and the development of Vienna in all possible manners."[12] This "development" alluded to the establishment of new, large commercial and industrial zones designed to underpin the city's economic growth. The Danube's regulation did not cause this industrial usage of the river: cities had long relied on the river and their own municipal waterways to provide industrial services—transportation, power, cleaning, cooling—for metallurgical works, mills, textile processing facilities, and other industries.[13] On the other hand, the invention of the steam engine and its early use in river navigation contributed to the river's industrialization.

Steam navigation encouraged and strengthened business connections to the river; both those trading finished goods and those relying on deliveries of raw materials. Steamboats could transport bulk goods more quickly, garnering a growing share of this trade. The Danube Steam Navigation Company moved quickly to build up infrastructure along the river to ensure that its transportation operations were ready for this boom in trade and industry. When the company built its shipyard at Óbuda in 1835, it was one of the largest industrial institutions in Hungary.[14] After two years, the company was able to stop buying steamboats from Britain and produce them domestically at its shipyard. Offering initial employment to 60 men, by 1841, the shipyard employed over 400.[15] This shipyard, and another near Vienna at Korneuburg "[became] of extreme importance and constitute[d] a vast industrial zone specialized in naval construction."[16] By 1879, the DDSG had 1,655 workers at their Óbudu shipyard and over 200 at each of their Korneuburg and Turn-Severin shipyards. The Korneuburg shipyard would eventually house 9,000 people at its expansive worker colony.[17] In the early decades, qualified British technicians received lucrative salaries to repatriate and work in the DDSG's shipyards or on ships as mechanics, often securing roundtrip passage for themselves and spouses.

Industrialization of the Danube followed regulation projects, as floodplains disappeared and new land adjacent to the river became available for factories, warehouses, and trade facilities. The so-called Danube City (Donaustadt) emerged as a new industrial zone in the Danube's former floodplains. While official positions praised the regulation's positive impact on trade at the imperial capital, members of the Lower Austrian Trade Association (Gewerbeverein) published an 1891 policy paper expressing disappointment that the Donaustadt's growth was less impressive than projected.[18] Nevertheless, Vienna's development was tied up in steam navigation and river regulation works. From 1830 to 1870, "the overall volume of wood, coal, food, feed and building materials brought into the city roughly tripled from 700 kt/year to almost 2.1 million tons/yr ... [while the] total amount of materials imported on the Danube more than doubled."[19]

Budapest's grand renewal works in the 1870s likewise benefited from the Danube's regulation and plans to develop its riverfront industries. In 1870, the Budapest Metropolitan Board of Public Works (Fővárosi munkák tanácsa) formed to design new elements for the city, which included plans to regulate the Danube, construct bridges, and beautify the quays and riverfronts. Contemporary documents recorded the board's desire to make Budapest "a really great city which, by its economic energy and renown, would attract tradesmen, shopkeepers, investors and nobility alike."[20] Ferenc Reitter, the first president of the board, placed the Danube at the heart of these plans. Besides major projects to regulate and embank the main arm, he also hoped to transform a swampy side arm of the Danube into a central shipping channel and promenade. He wrote, "The idea has been so seductive that I cannot tear myself away from it. . . . When I finally understood the influence that such a canal would have on shaping the future of Pest . . . it seemed to offer everything that we need."[21] Reitter's plans for the main bed were realized, but despite his best efforts to the contrary, the city chose to fill in the waters of the envisioned shipping canal and instead win land for Budapest's prominent Grand Boulevard (Nagykörút). Nevertheless, the idea of a network of canals around the city cropped up periodically in the coming decades.[22]

As governments and companies built up rail and river infrastructure, cities were incentivized to invest in better commercial hubs. Budapest and Vienna saw dozens of warehouses and steamboat landing places connected to customs' houses and rail stations.[23] In the early 1870s, the famous architect Miklós Ybl designed the new Customs House in Budapest, which four hundred to five hundred workers constructed between 1871 and 1874. The building had four tunnels connecting it to ports on the Danube. To more safely store goods transferred between rail and water, merchants and businesses were keen to have public and private warehouses built. This transfer and storage of goods also required equipment and buildings. Budapest boasted the second largest grain milling industry in the world after Minneapolis, Minnesota, and in the early 1880s, the Austrian architect Christian Ulrich designed a massive new grain elevator to serve as "monumental symbol" of this industry. The elevator featured a huge grain silo and could move grain to and from boats and railway wagons.[24]

Imperial and national authorities also increasingly fielded requests from cities that hoped to secure funds so that they could build transshipment hubs that would better integrate their rail and river commerce. Most of these cities used similar language about the benefit that such infrastructure would have on trade and on the empire's economy. The often repeated phrase that many cities deployed was that quays, warehouses, transshipment hubs and river regulation would together make these cities the center of east–west "world trade."[25] For imperial and national authorities, supporting municipal development in a manner that likewise promised to bolster the empire's standing in international affairs like commerce was a tantalizing prospect, and one that authorities undertook to the best of their financial abilities.

Looking at the nineteenth century, industrialization of the Danube appeared promising to politicians and businesses that sought economic vitality above all other measures of the empire's health. For these groups, trade, transportation, and industry were critical markers of this vitality. Commercial records and government statistics, along with lobbying efforts and petitions, reveal the hyper-focus on measuring and mobilizing data to support and justify hydraulic engineering works and industrial infrastructure. The twentieth century witnessed a whole new level of industrialization along the river from hydroelectric dams to heavy industry; most of which brought dire ecological consequences in their wake.[26] The successor states that pushed these projects, though different from the Habsburg Empire in many respects, pursued many of the same outcomes for industrializing the river: economic growth, political stability, and social cohesion. In this regard, industrialization and the optimism (and devastation) around it outlasted the empire and continued to shape cities up and down the Danube in the following century.

Healthy Bodies, Healthy Cities

Urban spaces grew as a direct result of industrialization and featured more and more smoke-belching factories, slaughterhouses, and other industries dumping effluent into urban streams or sinkholes; growing populations with limited sewage facilities and overpacked housing spread communicable diseases. These developments directly threatened public health and led technical and medical experts to imagine new designs for healthier cities and healthier citizens.[27] Beyond the more benevolent intentions, physical health began to play a prominent role in many nationalist historiographies, where the trope of citizens' health served as an allegory for the health of the "nation."[28] Healthy bodies indicated strong moral and national health, and vice versa. Such rhetoric emerged in nonnationalist, political discourse as well—for example, the concern that Socialist and Social Democratic parties placed on the health and morality of the proletariat classes.[29] Tied up in these political-philosophical discussions were the actual material conditions that people lived in, which affected their physical health. Especially in rapidly urbanizing cities, this question of living conditions—and the role of built and "natural" environments—weighed on city planners, politicians, and the public itself. In Vienna, for example, housing had become so expensive that homelessness represented a serious problem, with one satirical paper quipping that "for decades poor people had been using the riverbanks of the Danube as a substitute for cheap flats."[30]

The Danube became linked to discussions about urban health because of its historical role disposing human, domestic, and industrial waste. The Romans at Vindobona (Vienna) had developed a sophisticated system for flushing debris outside the city limits into the river. This practice halted with the empire's collapse, until Maria Theresa reintroduced a municipal sewage system to Vienna in the eighteenth century. Municipal ordinances indirectly encouraged the use of the river as a garbage dump. When new laws tried to stop residents from dumping waste in public areas, they simply jettisoned ordure and debris into either the Danube Canal or one of several urban brooks that flowed to the Danube.[31] The magistrate in Buda actually directed citizens to place garbage near the Danube's edge so that subsequent floods would carry it away.[32] One author noted with distaste that "on the banks of the Danube, the stench of garbage heaps and rotting material in the mud did not make for a pleasant walk."[33] Until microbes replaced miasmatic theories as the cause of communicable diseases, few understood that such practices were dangerous for the public's health.

Municipal authorities began to reconsider these practices as changing weather patterns and urban growth exacerbated the spread of diseases in the late eighteenth and the nineteenth centuries. Aggravated by heavier rainfall, greater runoff (from melting glaciers at the end of the Little Ice Age), and more open field cultivation, extraordinary flooding in the late eighteenth and early nineteenth centuries inundated cesspits, spilling people's waste and fecal matter into wells and other freshwater sources.[34] When global cholera outbreaks struck in the 1820s, none understood that it spread through feces-infected water. Flooding in 1830 therefore worsened Vienna's cholera outbreak, eventually leading to the deaths of two thousand people. Another cholera outbreak in 1872–73 was far more devastating and caused nearly half a million deaths in Hungary. While the greatest loss of life was along the Tisza River, in Pest, districts near the Danube where poorer and less-educated residents lived experienced higher fatality rates than did wealthier, educated residents.[35] Small cities like Esztergom experienced cholera outbreaks in 1866 and 1886, and the Melk city council fretted constantly that the local Danube arm would silt up and cause serious hygiene problems for its residents. Other illnesses were likewise prevalent in the river-rich empire. Heavy rains and extensive flooding in August 1869 combined with a bad harvest highlighted the danger of widespread stagnant water in Hungary, as 14 percent of its population contracted malaria.

Meant to help beautify cities and clear away deadly miasmas, a minor network of sewage canals emerged in the early half of the nineteenth century. In 1830, Vienna possessed approximately seventy miles of sewage canals. In response to the 1831 cholera outbreak, the city built a sewage canal (1836–39) adjacent to the Vienna River, which residents called the "cholera canal." By the 1840s, city authorities also began entombing the city's brooks in subterranean sewage canals. Palatine Joseph mandated underground canal construction in 1801 to rid Buda's streets of ordure. Unfortunately, when ice dams blocked the Danube's path in 1830 and 1838, streams and rivers flowing into the Danube backed up as well, causing Buda's sewage canals to overflow. Widespread contamination followed, demonstrating the vulnerability of a system without sluicegates or other protections against floodwater. By 1840, the Building Committee (Építési Bizottság) made plans to build sewage canals in Buda and Pest, over which the city would retain oversight, but the costs would be shared between municipal authorities and landowners receiving sewage connections on their properties. By 1860, Buda had sixteen miles of sewage canals and Pest had thirty-four.[36] These presaged large-scale canalization in the 1860s and 1870s but proved insufficient for growing urban needs. They also brought the river's hydrology to the forefront as authorities realized the danger that the unregulated river and unmonitored practices of dumping garbage and offal into rivers and brooks had on the river's ability to sufficiently remove waste.

Recurring epidemic outbreaks and social uprisings throughout the century convinced municipal, national, and imperial authorities to revise cities' "irrational" layouts and antiquated construction, which relied on unsophisticated methods of dumping refuse into the Danube. From the 1850s onward, a new epoch of city planning occurred in large cities, such as Paris's "Haussmannization," Vienna's "Founding Era" (Gründerzeit) in the late 1860s, and Budapest's renewal through the Metropolitan Board of Public Works (Fővárosi Közmunkák Tanácsa) in the 1870s. Engineers in the German Empire and Habsburg Empire influenced city planning with new engineering theories about improving water-city relations, particularly concerning how best to provide fresh water and discharge waste.[37] Their suggestions improved earlier sewage canals and reaffirmed Vienna's ordinance—implemented in 1867—to keep water supply and refuse canals separated and to bar cesspits altogether, provided that sewage canals existed nearby. Pest was canalized between 1869 and 1910, as was Buda between 1873 and 1914, though the 1873 financial crisis initially hampered this work.

The expense of sewage canals and freshwater pipes was often so prohibitive that smaller cities lacked the resources to follow suit. Eventually, engineers made technical plans available to cities, even recommending different types of sewage systems based on the city's particular geography and financial situation. Pozsony/Pressburg and Szeged started canalization in the 1870s, by the 1890s, towns like Arad, Győr, Miskolc, and Kassa had begun work on their sewage and waterworks systems centered around their respective rivers (Maros, Rába, Sájo, and Hernád). By the turn of the century, 36 percent of the cities in the empire's Austrian half had some form of sewage system, though this figure was only 13 percent of the cities in the empire's Hungarian half.[38] As more complex canalization projects flushed ever greater volumes of waste into the Danube, engineers looked to examples in London and the Netherlands for treating water and extracting organic solids valuable for agriculture.[39]

While sewage and industrial effluent remained tied to the river for removal, authorities began to emphasize the need for healthier public practices regarding garbage disposal. Imperial and provincial "water rights laws" in Austria in 1869–70 delineated the proper usage of public water and promised to prosecute any infractions, though these regulations were not always successful in changing practices. In 1874, the national authorities in Budapest passed a law on public health underscoring their responsibility for maintaining sanitary conditions and outlining punishments for infractions.[40] Hungary's 1885 Water Law likewise prohibited the dumping of waste into waterways before treating it, but in practice, according to Endre Juhász, "getting rid of sewage as soon as possible was considered more important than protecting the water of the recipients."[41] In 1879, the St. Pölten district head excoriated the pollution of water sources in the *Amts-Blatt*, informing residents: "In recent times, the most unsanitary conditions have been observed and natural watercourses are generally seen as misused, with the introduction of all kinds of waste and pollution of everyday life, such as waste from industry, humans, and partially from cattle, so that streams and smaller rivers, rather than bringing the beneficial influence of a flowing, pure body of water, have become places of putrefaction, which pollute the air and contaminate drinking water."[42] In smaller communities, the local councils voted to erect signs forbidding such practices. In Melk, council members even voted to close off the pathway leading to the river to prevent people from dumping garbage along its banks. The presence of continued ordinances and regulations in the last quarter of the century suggests that the river remained a temptingly convenient place to get rid of undesirable junk and waste.

City residents also grew healthier thanks to a shift away from drinking Danube water. Emperor Ferdinand had overseen construction of a water supply system in the 1830s that provided Vienna with water from the Danube Canal. It did not properly clean the water, and disease outbreaks were still common. By 1864, the city council approved plans to build aqueducts from the surrounding mountain ranges to provide Vienna with fresh spring water. Franz Joseph presided over the opening ceremonies in 1873 that coincided with its hosting of the World's Fair. An article about Vienna's new sanitary conditions praised this project, arguing that since it opened "dysentery has now become quite unknown . . . typhoid fever has also well-nigh disappeared," though it cautioned that when the reservoirs froze in 1877 and residents again received Danube water, typhoid infection rates ballooned to 21.5 per 10,000 people compared to only 3.8 per 10,000 in districts where Danube water was not distributed. It concluded, "These statistics should be committed to memory in every municipality, especially by the authorities of those that are being supplied with unfiltered, filthy river water which receives and dilutes the offal of communities and again distributes them whence they came."[43] While Vienna's proximity to mountains made such alpine water sources possible, Budapest's lower geography led the city to exploit its freshwater aquifers instead, an example followed by many cities in the empire's low-lying plains in the east.

Broadly speaking, waterworks continued to expand throughout the empire's municipalities: in 1899, 53 percent of cities in the empire's Austrian half were supplied with fresh water and 28 percent of Hungary's, though these masked large regional differences—75 percent of Bohemian cities had freshwater pipes but only 25 percent of Galicia and Bukovina's cities did.[44] By 1908, statistics indicated that 25 of Hungary's 28 cities "with legislative authority" had pipelines of some kind.[45] It was only World War I that eventually halted progress in Hungary's public works.

Riverbanks as Recreation

In 1807, the *Wiener Zeitung* published a circular from the Lower Austrian authorities reminding the residents of Vienna that due to the rapid pace of the Danube, public swimming was forbidden and rule-breakers would be "brought to the nearest police station and punished." Striking a more conciliatory tone, the circular then assured people that for their sanitation and well-being, there were designated areas along the river that were open to the public for the purpose of bathing.[46] As debates about the river's usage oscillated between protecting the water quality and using the Danube for waste disposal, this circular serves as a reminder that the river was not just a commercial or industrial waterway but also benefited the public in daily, individual ways: recreation.

By the late eighteenth century, river baths had begun greatly enriching the river's appeal to the public. In 1799, Franz II permitted the Lower Austrian government to establish two public baths, one for men and one for women, in the Tabor Arm of the Danube. Donations from "patriotic" individuals funded the baths. In 1806, the Private Humanities Association set up a large bathhouse on the Moldau in Prague.

After it wore down from wear and tear, various associations adopted the cause and helped rebuild the facilities with several bathing pools available for the paying public. Later, cheaply run *Flussbäder* or "river baths" appeared during the *Vormärz* period, which municipal authorities set up to provide cheap bathing opportunities for the poor, in a distinct reversal of the centuries-long ban on public bathing in the Danube.[47] To ensure that public modesty remained intact, officials monitored access to designated bathing spots on the river.

These swim facilities were found throughout the empire and set up to cool and clean the public. "Ship baths," wooden bathing houses floating on pontoons and anchored to a section of the riverbank, also provided services for up to 1,000 bathers a day during the summer.[48] In 1858, a private bathing facility opened in Melk along the Danube sidearm flowing past the city. By 1863, the Korányi József pool was opened on Esztergom's Small Danube channel, and the pool operated until the turn of the century. At the end of the nineteenth century, there were five bathhouses along the Danube Canal, which became even more popular when new sewer lines made the canal suitably clean for swimmers.[49] The Public Bathhouse (Städtische Badeanstalt) on the Danube's main arm had a massive swimming pool, bathing basins for "nonswimmers" as well as private baths, all of which could accommodate 1,200 guests. The bathhouse's café-restaurant also provided an attraction for guests, as it afforded diners a view over the newly regulated river. By the late nineteenth century, the three public floating pools on the Danube at Vienna registered 50,000 yearly visitors on average, and Budapest's three public pools on the Danube had over 120,000 visitors.[50] Even during World War I, people of all stripes escaped the summer heat down by the canal. According to a *Neue Freie Presse* article in August 1915: "Making your way to the Danube Canal on a sultry late afternoon, you will become aware of the metropolitan summer. On the canal's embankments soldiers and convalescents enjoy their modest recovery. Poor people also pursue freshness here. Their miserable-looking shaggy dogs bathe, they take sunbaths free of charge or sit on their shores with their trousers tucked up, leaving their feet in the water. Soon enough they are scared off by guards, evidence of an illegal footbath without concession or certificate of permission."[51]

Beyond the river's appeal for swimming and bathing, an 1883 English edition of Badaeker's popular travel guide made it clear that the riverside was rife with amusement for people of all classes. The Danube's regulation had, in many instances, created these new spaces for people to enjoy the river. Already in 1766, Joseph II had opened the imperial hunting grounds at the Prater, an immense island between two arms of the Danube, to the public. From 1781 to 1783, Isidor Canevale built the Lusthaus at one end of the Prater Allee, which served as the site for many imperial and local celebrations, including as, chapter 1 indicated, the location where one of the first steamboats, *Franz I*, docked for the Viennese to inspect and fete in 1823. In the nineteenth century, the Prater was a meeting ground for the bourgeoisie, especially after the main thoroughfare leading to the park, Jägerzeile, became one of the most fashionable and desired locations to live in Vienna.[52] By the late nineteenth century, as the Badaeker guide explains, the Prater was likewise "the favourite haunt of the lower classes, especially on Sunday and holiday afternoons" because of the "attractions adapted to their tastes."[53] As discussed in chapter 3, the 1873 World's Fair took place on the grounds of the Prater.

New quay promenades and even bridges on the water provided opportunities for the public to enjoy the river views. While Budapest's quay construction belatedly began in the 1860s, political leadership required the entire twenty-four-kilometer site to have a uniform style, which incidentally provided a more aesthetic perspective. When suggesting designs for a bridge connecting Buda and Pest with Margaret Island in 1832, Pál Vásárhelyi had envisioned lookout points on the bridge to enjoy the view of the twin cities, and when the Margaret Bridge was later completed in 1876, it provided pedestrians with a panorama of the city "hitherto only experienced by ship passengers."[54] In 1864, the Large Danube promenade in Esztergom opened, helped with the support of the town's mayor, Ágost Forgách. The promenade enabled pedestrians to enjoy the beautiful sight of the new Esztergom Basilica—the seat of the Catholic Church in Hungary. Franz Joseph attended its commemoration in August 1856. Linz likewise created new walkways along the Danube. The local beautification club installed benches for walkers to sit on, and planted gardens adjacent to the river, in particular near the new DDSG headquarters near the quay. In April 1899, Imre Miller even opened a few confectioneries along the Danube at Kioszkot to feed passersby. By 1913, Esztergom's Promenade Beautification Club (Sétahelyszépítő Egylet) adopted the Little Danube's bank as their next project to enhance.

Figure 5.1. Rowing Regatta on the Danube Canal, 1903 © Brandstätter Verlag.

The proliferation of educational, recreational, and political associations connecting communities across the empire also opened the venue for organized socialization along and on the Danube. Rowing, for example, became a popular activity on the river, one that both entertained local communities and brought together teams from different clubs around the empire in friendly competitions. István Széchenyi, as an Anglophone, promoted the British activity in the 1840s with the club "Csolnakda" for rowers. Participation grew after April 1861, when twenty-nine individuals, twenty-three of whom were counts or barons, founded the Budapest Boat Club (Budapesti Hajós Egylet). Within two months, the new club had grown to fifty-two members.[55] In 1862, the Pest Rowing and Sailing Circle brought about the establishment of a series of other rowing clubs. In April 1863, the first modern rowing competition was organized on the Pozsony/Pressburg stretch of the Danube, and by June that year, Pál Pakson had organized the Tolna County Rowing Club.[56] That same year, the First Viennese Rowing Club organized a team for youth who had been rowing around the Danube wetlands (*Auen*) after ostensibly bringing the activity back from their trip to Britain. In the 1870s, additional cities in Austria organized rowing clubs, such as Linz and Stein on the Danube and Klagenfurt on Wörth Lake, and by 1891, there were already fifteen clubs in Vienna alone. To encourage rigorous training, the Viennese (1874) and later the Corinthian Regatta Committees were established to ensure more organized and regular regattas (fig. 5.1). In the 1900 Olympic Games, rowing joined the summer events, and an Austrian, Alfred Heinrich, took part.

The clubs established themselves and their boathouses along the Danube, and they also participated in the social life of the cities, hosting parties and events during the year. In January 1871, the Budapest literary journal *Fővárosi Lapok* advertised the annual carnival lineup, announcing that the National Boat Club and the Duna Rowing Club were joining together to put on the Boat Ball (*hajósbál*), which the journal claimed was "always very fancy."[57] The Linz-based Viking rowing club invited guests to celebrate with the team before rowing meets. During the winter, it petitioned the city council to set up an ice-skating rink near the river, which was such a success that they repeated it each winter thereafter. When boat traffic decreased on Wiener Neustadt Canal, the waterway evolved into a site of recreation. Skating on the frozen waterways in Vienna—previously forbidden by municipal decrees—was allowed by the authorities, which designated particular spaces, like the Danube Canal, as open to winter activity. Over the course of the century, the canal's informal, natural ice rink became "institutionalized" as an association took over the construction and maintenance of an increasingly sophisticated rink. The new space provided Vienna's bourgeoisie a place to ice skate and occasionally brush shoulders with Emperor Franz Joseph I, who held social galas on the ice rink.[58] After the onset of World War I, many clubs on the river disappeared due to members being mobilized and the military requisitioning boathouses for supply purposes.

Case Studies of Linz and Győr

As cities changed, so too did the Danube. New commercial-industrial, sanitary, and recreational spaces offered urban residents the opportunity to interact with each other and use the river for everything from workplaces to leisure spaces. The empire's citizens were active participants in the creation of these spaces. Looking at the examples of Linz and Győr, the capitals of Upper Austria and Győr County, respectively, we see how local actors played a key role in navigating and shaping the transformation of the Danube. Their interactions with governments, clubs, and businesses demonstrated how negotiations and civic engagement empowered them to protect their local interests and initiatives. Their flexibility strengthened political, economic, and social institutions both locally and imperially, making riparian communities a crit-

Act Locally, Think Imperially

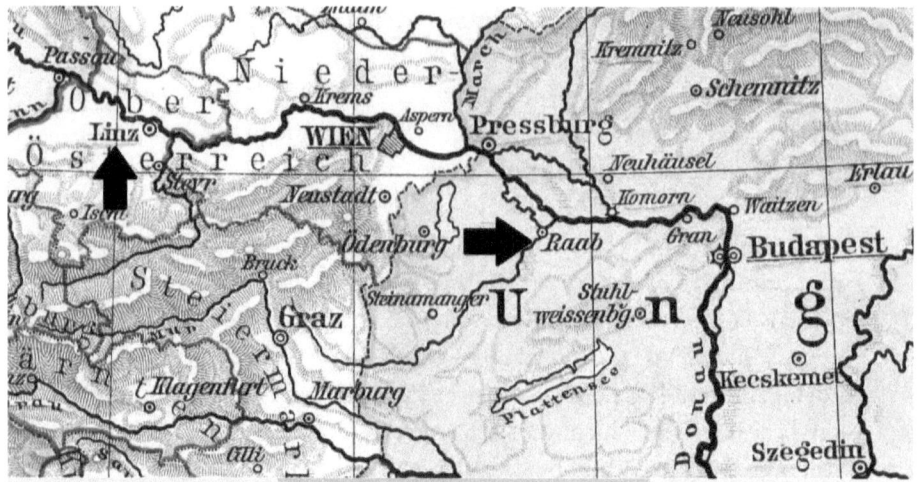

Figure 5.2. The Positions of Linz and Győr (Raab) in relation to Vienna (WIEN), Bratislava (Pressburg), and Budapest. *Source:* "Österreichisch-ungarische Monarchie und die Schweiz, Staatenkarte," *Lange-Diercke Volkschulatlas* (Braunschweig: Westermann Verlag, 1898–1904) © Westermann Gruppe.

Figure 5.3. Linz, Strasser Island, and the Fabrikarm in the Late Nineteenth Century, 1890. *Source:* Hans Wöhrl, Regulation of the Danube Arm at Strasser Island. Photograph, 1890. Nordico Stadtmuseum, Linz, Austria. Archival number: NA-021457.

Act Locally, Think Imperially

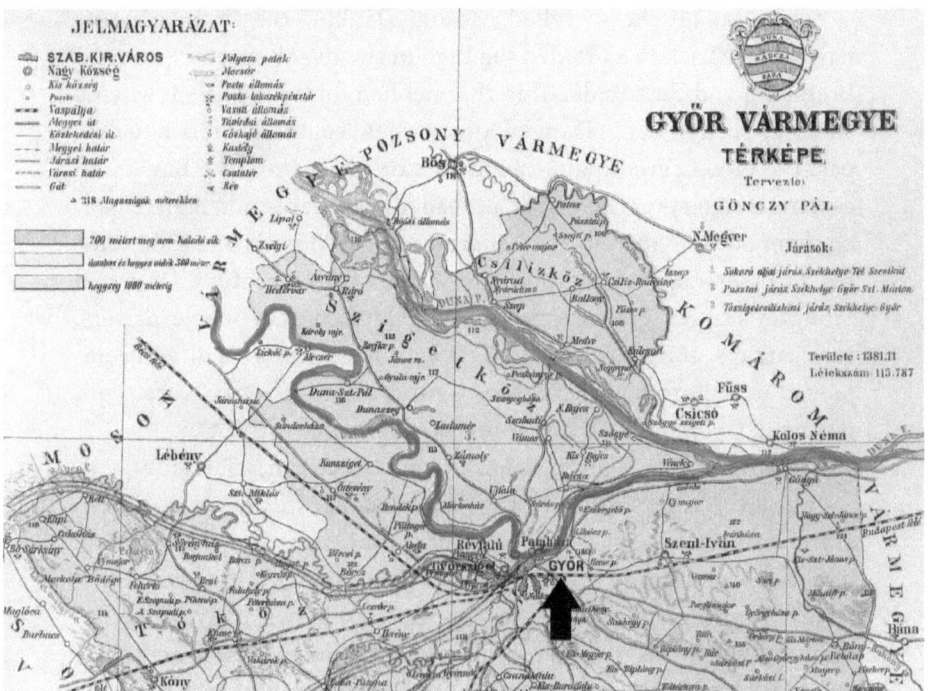

Figure 5.4. Historical Map of Győr on Mosoni Danube with Main Bed. *Source:* Manó Kogutowicz, *Győr vármegye térképe* (Budapest: Hölzel és Társa Magyar Földrajzi Intézete, 1891).

ical component of the empire's internal cohesion and ensuring their resilience in the face of innumerable changes to their local environments.

Both at Linz and at Győr, the Danube sat at the center of each city's historical prosperity and development (fig. 5.2). In Linz, a large flood in 1572 had washed through town and created a new channel adjacent to the city. This so-called *Fabrikarm* (factory arm) facilitated ship traffic to and from the city and provided a site for early industrial development (fig. 5.3). The Linz Municipal Brewery (1590) established itself here, as did a textile factory (1672), which at its height employed over forty-nine thousand people in the surrounding countryside. Several mills on the branch processed the wood that lumberjacks sent downstream from throughout Upper Austria.[59] The river's kinetic energy powered several mills both on the main channel and the side arm.[60] Regional and international merchants visited its biannual markets, and its location made it an important transit point for merchants traveling north to south overland and east to west on the river.[61]

165

Győr sat on the so-called Mosoni Danube, the Danube's only navigable branch in a braided segment of the river between Pozsony/Pressburg and Pest-Buda. This channel had formed in 1653, when a silted-up section of the Danube forced a new, southerly branch to flow past the city. Crews towing shipments of wheat from the Banat and eastern Hungary reached Győr, unloaded their wares, and then reloaded them onto smaller boats to continue the final stretch of their journey to Vienna. By the beginning of the nineteenth century, Győr had Hungary's largest grain export market.[62] Hundreds of people, drivers, sack carriers, and sack rental and repair businesses, were directly employed in the grain trade.[63] Some figures estimated that every tenth person in Győr made a living from this commerce (fig. 5.4).

Steam navigation promised to expand the commercial ties of both Linz and Győr. In spring 1831, Győr welcomed steamboat traffic every few weeks once the DDSG began running the *Franz I* regularly between Vienna and Pest. In 1832, the ship expanded its route downstream, which resulted in a lively goods trade between Győr and Zemun, a town across the river from Belgrade.[64] In an 1833 letter to the palatine, the Hungarian aristocrat István Széchenyi mentioned that "passenger traffic from Győr to Semlin and back through Pest is now getting brisk," leading to a hefty profit for the DDSG.[65] The river's traffic was so important for residents that many built granaries instead of houses, particularly near the riverfront, which had long held the city's highest property values due to its proximity to river commerce. By mid-century there were 147 granaries in town, and the Danube riverfront was completely built up.

Even before steamboats arrived in Linz, the city's residents were reading about this novel and exciting means of travel. One local paper published a series of articles romantically describing a trip down the river, highlighting the historical and natural wonders that travelers would experience. When steamboats finally arrived in 1837, Linz's authorities placed the steamboat landing place prominently next to the city's bridge spanning the Danube and signaled every boat's arrival with cannon fire.[66] Their landing place on the *Fabrikarm* shared the space with mills, factories, swimming facilities, and barracks—a combination of different interests that eventually required several rounds of negotiations to ensure fair access to the river's usage.

As steam navigation expanded throughout the Danube, commercial groups throughout the empire called for the Danube's regulation

Act Locally, Think Imperially

to ensure sufficient depth of the river unencumbered by sandbanks or shoals. Győr's merchants regarded these plans with apprehension. When the Lieutenancy Council, the Habsburg advisory body in Hungary, mandated that the Danube's main Komárom branch be made navigable, it threatened to divert traffic away from Győr by bypassing the narrower Mosoni channel altogether. Although the Hungarian authorities had only partially regulated this stretch of the Danube by the mid-1840s, the Danube Steam Navigation Company already considered skipping Győr and making Gönyű (on the main branch) its scheduled stop on its daily trip between Vienna and Pest-Buda. This plan sent Győr's municipal authorities scurrying to plead with the DDSG's administration to keep it on the schedule.[67] Writing to Count Széchenyi to complain, the DDSG head Joseph Voigt declared that poor conditions in the "Győr channel" were slowing river traffic and he suggested that Széchenyi bring this up with Győr's commercial parties.[68]

Over the next decade, the slow progress on the regulation of the Danube's main bed ensured that the Mosoni Danube's traffic remained steady. Győr's city council did little to dredge or regulate the river until the 1850s due to the continued strength of grain transports arriving in town and the absence of Hungarian rails undermining its river traffic. Even in 1856, the Pest-based newspaper *Pesti Hírlap* reported that Győr's grain traffic was still three times that of Pest's.[69] However, the next few years witnessed a distinct downturn in the river's commerce, as the spread of railways soon threatened to derail Danube navigation at Győr. In 1855, the city's first rail line arrived, connecting it to Bruck, and by the following August of 1856, it extended to Újszőny (modern-day Komárom). Once the rail line opened and connected Győr to markets in the west, wheat arriving on the Danube could simply transfer to rails heading to the Vienna or send it on the Südbahn down to Trieste, rather than continue up the Mosoni Danube.

With railways expanding, business interests stepped in to advocate for the river's maintenance. In 1856, a group of merchants founded the Commercial Guild to promote, among other interests, the Danube grain trade. In fall 1857, the guild wrote to the mayor to decry the unmaintained Danube near the city. The guild claimed the river's conditions were causing low water levels and preventing larger ships from traversing up the Mosoni Danube to Győr from Gönyű. To continue transporting shipments, merchants had to pay to unload and reload goods from larger to smaller ships, which supposedly cost four hundred florins per boat in delays and labor costs. Unsurprisingly, the guild feared that these costs would harm business.[70] The guild believed that commerce's importance for the town and region meant that the Danube's maintenance was good "not only for the city's residents but for the entire public as well."[71] Nevertheless, by December 1858, the local paper *Győri Közlöny* was bemoaning the fact that due to growing rail competition, the DDSG was halting its Little Danube (Mosoni branch) traffic.[72]

The Danube Steam Navigation Company pleasantly surprised residents in Győr by not only keeping its stopover there but also extensively advertising its routes and rates in the local newspaper, the *Győri Közlöny*. Residents did take issue with the fact that boats leaving Pest arrived between two and three o'clock in the morning. In April 1859, the *Győri Közlöny* aired residents' grievances, arguing that such arrival times meant that passengers had to stumble home in the dark from the steamboat station.[73] The DDSG directed its ships to depart a few hours earlier to address these concerns, but residents complained next that "arriving at midnight was not much better."

The DDSG employed additional strategies to ingratiate itself to the public and maintain smooth operations for its river traffic. For decades, the company had spent its own funds to clear obstacles and hindrances to navigation on the Danube, a stipulation of its 1846 subsidy agreement with the imperial authorities. When residents of Győr fretted that Danube traffic would pass them by, the company launched a survey of the Mosoni Danube in 1860 to better assess how to maintain the channel's navigability, and therefore profitability, for the town.[74] These insights enabled the city council to earmark funds to cover the costs of dredging. When a flood struck the following year and reconstructions costs strained the town's finances, the DDSG offered to help with the dredging costs.[75] Besides ensuring the Danube's navigability, this move likely sought to shore up local support for the DDSG, as the new Royal Bavarian Steam Navigation Company had just started advertising its freight services in the *Győri Közlöny*.[76]

For commercial representatives at Linz, the Danube's unregulated state likewise signified lost business opportunities. In 1852, the Chamber of Commerce complained to local authorities that the river prevented larger steamboats from safely or easily reaching the city.[77] Better conditions would be needed to encourage trade and to support the town's growing shipbuilding industry. A new shipyard had sprung up on the *Fabrikarm* in 1840, established by a local shipbuilder, Ignatz Mayer. In 1854, Mayer employed 200 men building and repairing steamboats, which grew to 550 by 1869. The shipyard provided services for a transnational clientele: servicing boats for the DDSG and its Bavarian-Württemberg competitors and even constructing boats for a Hungarian steamboat company in 1867. Regulating the Danube would be critical to its success. The Chamber of Commerce also had an eye to expand steam navigation to Vienna. The newly completed Südbahn railway connected Vienna (and the Danube) to the port city of Trieste on the Adriatic Sea. Members hoped that this new link to southern markets and Mediterranean trade would mean that Linz would enjoy a large explosion of steamboat traffic coming up the Danube as a result. Instead, the newly extended Westbahn connected Linz directly to Vienna in 1858, and a gradual decrease in Danube passenger traffic resulted, a trend that would not reverse until the 1890s with the construction of a new transshipment hub.

With fears that expanding railways would further erode Danube traffic, commercial groups in Győr felt the need to take more active measures to protect and promote the town's trade on the Danube. These concerns seemed even more justified as the milling industry downstream in Pest appeared poised to eclipse Győr in the 1850s.[78] In 1858, 800 ships had transported nearly 700,000 tons of grain to Győr, which fell to 320,000 tons on 400 ships in 1867. In 1865, the local paper published a series from the newspaper *Új Korszak* (New Era) advocating for the establishment of a specifically Hungarian steam navigation company. Regional merchants took a decidedly more local route, setting up the Győr Steam Navigation Company to combat the DDSG's unfavorable shipping rates, promote commerce, and enable the "reasonable transport of wares via steamboat on the Danube and its tributaries."[79]

As discussed in chapter 2, such regional companies often had greater success than their national counterparts in carving out important niches commercial markets. The company had 5 freight steamers, 21 tugboats, and an additional 31 wooden boats for trade. This fleet paled in comparison to the DDSG, which in 1860 had 95 paddle steamers, 24 screw propellers, 381 iron goods transports, 26 iron cattle transports, 42 iron coal ships, 17 iron standing ships (*Stehschiffe*), and another 15 miscellaneous ships. However, it was one of the first non-DDSG Danube steam navigation companies in the empire. Pragmatically negotiating with the Danube Steam Navigation Company to divvy up the Vienna-bound grain trade, Győr's steamboat company had relatively steady success maintaining its shipping numbers. Residents likewise felt buoyed by the announcement in 1868 that the national government in Pest-Buda planned to regulate the Upper Danube and its tributaries, including the Small Danube from Győr to Gönyü, though regulation work on this local stretch only began in 1885. With the newly formed national government in Hungary after the 1867 Ausgleich, the Commercial Guild reassembled its members to build on these opportunities to boost commercial performance on the Danube.

While commercial interests grappled with the implications of steam navigation and railways, the unregulated Danube remained a considerable source of unease for residents of Linz and Győr due to its perennial flooding. Not only did an 1876 ice jam on the Middle Danube cause massive flooding along its banks but the blocked river also flooded its tributaries, including the Rába flowing through Győr. Another flood on the Danube in 1883 led the county's head official (*alispán*) to declare a state of emergency, given that ten thousand people remained without homes.[80] The situation on the Danube and the underregulated Rába River appeared so dire that the Győri Lloyd (business center) penned a letter to the Royal Hungarian transportation minister in February 1883. Its complaints not only cited the decline in commercial ties—due to competition with Vienna and Pest and those cities' favorable commercial arrangements—but also decried the existing hazards that the unregulated Danube and Rába Rivers presented to the town's residents.[81]

To keep its residents safe, Győr's city council examined all technical and political partnerships that would aid its efforts. In 1884, it established a "flood prevention" committee to examine and suggest improvements to local embankments erected to defend new industries and city property in low-lying quarters. To help fund local projects, it turned to the national government. Frequent flooding from 1876 to 1883 and subsequent reconstruction had placed heavy tax burdens on the residents. The council petitioned the national authorities to help share some of the financial burdens expected from constructing embankments, raising bridges so that large steamboats could pass beneath them, and excavating sewage canals.[82] Demonstrating its commitment to these public works, the city council passed Law XV in 1885, providing it the authority to divert funds to implement flood protection measures.[83]

Working with the flood prevention committee, the mayor began the legal and financial process of acquiring private land near the Danube and Rába Rivers for embankment companies to build along. Not all individuals and companies living on the river or using its waters supported this process. For landowners near the river, their proximity to the river provided direct access to the river's trade, which inflated the value of their property. They therefore fought local authorities' efforts to appropriate it as "public land" for the purpose of erecting embankments. Certain professions also looked on nervously as changes to the river were proposed. Water mills caused siltation, which in turn exacerbated flooding, but millers consistently thwarted organized efforts to change their access to and use of the river in the name of river regulation.[84] In spring 1886, the city council invoked Law XV to justify these acquisitions. The city spent the next few years liaising with engineers and regulation companies in town to negotiate land acquisitions and embankment specifications. Municipal engineers provided detailed plans for where and how to regulate the city's rivers to best protect the city's residents and their property.[85] Such local initiatives reveal the concerns motivating members of the public and elected officials in grappling with the unregulated Danube.

While the residents of Győr looked for a solution to their flooding problem, flooding in Linz revealed the rippling consequences that these natural disasters had on a community. In 1864, a massive flood tore a steamboat loose from its mooring in Linz and crashed it into several of the pylons supporting Linz's wooden bridge spanning the river. Two yokes collapsed, closing the Danube to river traffic and severing Linz's connection with its neighbors across the river. The city council quickly agreed to rebuild the bridge out of iron, and in 1872 the new bridge opened to great fanfare. Only after the bridge's construction was complete, however, did the community start to feel the adverse effects of the river's new morphology.

While the bridge was under construction, the Linz city council had set out an ambitious plan to excavate a citywide sewage system. Drawn up at a time when the *Fabrikarm* still flowed strongly through the edge of the city, the plan directed the city's waste into the channel's lower end, where it could be washed away into the Danube. Once the bridge opened in 1872, however, its large new pylons diverted the

Figure 5.5. The Swimming Facilities in the Fabrikarm, circa 1880. *Source:* Unknown photographer. Strasser Island, Swim School. Nordico Stadtmuseum, Linz, Austria. Archival number: NWB-000249.

Danube away from the side channel. As the water speeds in the *Fabrikarm* slowed, the channel became stagnant with unwashed debris and wastes.[86] Members of the public compounded the problem by using the channel as a garbage dump. The city council reacted by passing a law forbidding the jettisoning of garbage into the river. This was so ineffective that a few years later they passed another law setting up garbage collection points next to the river and arranging carts for its regular removal.[87]

The declining conditions in the *Fabrikarm* brought together public and private interests to address these issues. One of the first parties to reach out for help was the director of Linz's bathing facility (fig. 5.5), who asked the Lower Austrian Danube Regulation Commission (DRC) to borrow an excavator to clear out the side channel.[88] The DRC demurred, claiming that it was not a provincial matter. After the director requested an excavator from the DDSG, the company only agreed to sell rather than loan one. Even the Imperial-Royal Interior Ministry claimed that the channel's regulation, unlike the Danube in general, was a purely municipal matter.[89] As the channel slowly silted up, in February 1882, the steamboat company requested exclusive use of the banks above the bridge for its landing place. Surprisingly, the city council denied this request in their next meeting in March, claiming that practices along the river needed to serve the "general good" and not just "one company's interest."[90] Taking matters into its own hands, the bathhouse moved its location to escape the insufficient water levels and increasingly unsanitary conditions. Unfortunately, the temporary facilities clashed with the newly established rowing club, which wanted access to that stretch of the riverbank for its practices. Facing declining operations, numerous businesses such as the tobacco factory, various mills, and other establishments mourned a lack of solutions to the channel problem.[91] The engineer Arthur Oelwein suggested that the only way to improve the channel's conditions would be to modify the whole main bed of the river. [92]

Although regulating the main bed of the Danube presented a much larger task, it also opened the opportunity for stakeholders to overcome opposition to purely "local" affairs and reinitiate negotiations with municipal, regional, and imperial authorities. Seeking to convince the imperial government to back these plans, representatives from Linz and Upper Austria's legislative bodies enlisted the help of the Danube Association (Donauverein). The association has been established in 1879 for precisely this task of lobbying the government and mobilizing public opinion for such projects.[93] The president of the Upper Austrian Chamber of Commerce, Johann Wimhölzel, had already addressed the Donauverein general assembly in 1882, passionately arguing against the Imperial Diet's opinion that the *Fabrikarm*'s siltation was an "internal matter" for Linz and urging the association to use its burgeoning influence to petition the imperial state on the city's behalf.[94] With its connections throughout the empire and backing from patrons in the Habsburg family, the Danube Association seemed to Wimhölzel and others well positioned to draw the imperial state's attention to Linz's affairs.

In June 1884, the Danube Association organized a "study trip" of the Danube, which stopped in Linz on its way from Passau to Vienna to see conditions in the *Fabrikarm* and discuss possible solutions. The participants were a diverse group. Representatives from several imperial ministries attended, as did local and regional government office heads and members from a dozen riparian communities, engineers from various cities, and members of numerous commercial and technical associations as well as those of steamboat companies in Austria and Hungary.[95] Linz residents festively decorated the riverbanks to greet the trip's participants.[96] The Donauverein held a meeting in the town's large ballroom in town.

Thanks to these creative maneuvers, negotiations on the Danube's regulation started again and revolved around concerns that public and private groups enjoyed common access to and usage of the Danube. As was typical for such public works, negotiators also attempted to hammer out a fair share of funding that each of the interested parties should bear. In late 1886, the city council was still debating how regulation plans would promote commercial interests, help the Bath House relocate, and end the channel's sewage problem, the latter two of which were important for the municipal population's health. The council also deliberated about how to compensate businesses in the *Fabrikarm* that were affected by any works undertaken by the city. The Imperial-Royal War Ministry's Marine Section offered to help pay for sewage canals near their barracks, which were located on Strasser Island across the *Fabrikarm* from Linz. In January 1887, the Upper Austrian Provincial Diet (Landtag) approved the War Ministry's contribution amount for the project on the condition that all types of boats—including rafts and galleys—would have equal access to the Danube.[97] By July 1889, the imperial-royal Upper Austrian governor's office approved the first payment installment and eventually in spring 1890, the Danube's regulation began.

Negotiations around the Danube's regulation promised to improve conditions for local businesses and offered the tantalizing prospect of making Linz a critical hub of transimperial trade. The city council encouraged these visions by decreeing that it would likewise set up a transshipment hub (*Umschlagplatz*), where ships could unload goods directly at the new train line, connecting the river traffic to the empire's railway network.[98] Agreeing that a transit hub would be important for Linz's development, the imperial authorities permitted the city council to set up a commission in 1884, which worked with local industries and military officials to ascertain how best to approach the hub's construction.[99]

In the early days of the planning phase, this commission emphasized the need to focus on public concerns and interests. In one of its first pronouncements, the commission stressed that planners had to

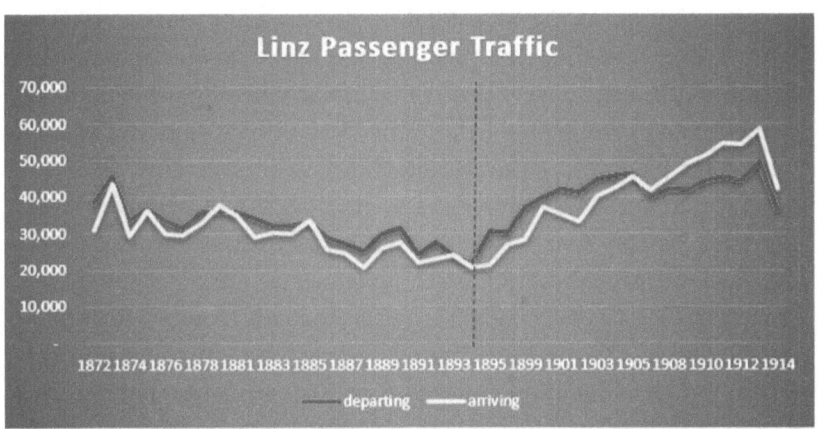

Figure 5.6. Linz Passenger Traffic, 1872–1914. *Source:* Annual reports (*Rechenschaftberichte*) published by the Linz City Council.

Figure 5.7. Linz Freight Traffic, 1872–1914. *Source:* Annual reports (*Rechenschaftberichte*) published by the Linz City Council.

consider the "general public interest and navigation safety" regarding regulation works before any plans for the hub's design could advance. President Wimhölzel of the Chamber of Commerce reminded members of the city council that in the interest of public health, any plans had to address the sewage problem bedeviling the *Fabrikarm*.[100] Likewise, when the commission broached the idea of closing off the *Fabrikarm* permanently, business leaders who relied on the channel responded that their livelihoods would end unless the city paid them an indemnity or could guarantee their relocation to another spot on the Danube.[101]

Negotiations on the transit hub stalled for a few years, until February 1886, when the city council decided to deploy delegates on its behalf to drum up support at imperial and international venues. It allocated funds to enlist the help of the Donauverein to present Linz's transit hub plans at the annual Inland Navigation Congress (Binnenschiffahrts-Congress) in June. It also charged its two delegates at the Imperial Diet in Vienna to bring up plans again during the next legislative session. The next month, Linz's delegate Hermann Vielguth gave a rousing speech on the Danube's importance to the House of Representatives, hyping the role that a transit hub would serve in uniting various commercial interests in the empire. Vielguth requested imperial funding for the Danube's regulation and the city's transit hub, which he opined had strong regional support among many governmental, commercial, and navigational groups and was likewise crucial for the "empire's unity." Pieter M. Judson has indicated that such rhetoric was common from provincial representatives seeking imperial support for infrastructure projects and investment.[102] Concluding his speech, Vielguth again accentuated the imperial significance of the Danube and the benefits accrued by investing in it, stating it is "the only great natural waterway, which runs through our entire monarchy from the west to the east, which connects the imperial capital with the capital cities of the other half of the monarchy."[103] The Diet enthusiastically welcomed Vielguth's speech and by May, the body had approved regulation and transit hub funding. In 1894, the new transit hub opened and the DDSG's steam traffic—both passenger and freight—increased markedly afterward, reversing a decades-long decline in passenger traffic (figs. 5.6 and 5.7).

By 1896, the new hub had attracted the attention of many businesses, locally and throughout the empire, looking to capitalize on the more profitable shipping conditions on the Danube (fig. 5.8). When the city council suggested opening a public warehouse at the hub, letters flooded the mayor's office in support of it. Local agricultural companies, iron and weapons firms, numerous steamboat companies from Bavaria, Austria, and Hungary, and even a milling company from Budapest all inquired about the space and conditions.[104] In July 1897, rail and river shipping companies, fruit merchants, millers, and representatives from the lumber industry all met in Linz to discuss details for the warehouse's construction.

Győr's investment in commercial infrastructure on the Danube

Act Locally, Think Imperially

Figure 5.8. Linz's Transshipment Hub (center buildings), Danube Steam Navigation Company office (far right), and Danube Steamboat Traffic. *Source:* Unknown photographer, *Anlegestelle an der Unteren Donaulände*, 1903. Nordico Stadtmuseum, Linz, Austria. Archival number: NA-051009.

also helped stave off economic decline. In 1887, the city completed new facilities to transfer goods between its river and rail traffic, taking advantage of the traffic boom that the relatively new Ebenfurt-Győr line brought to the city. As the minister of Public Works and Communication, Gábor Baross also oversaw regulation work on the Danube in the late 1880s, including on the Mosoni Danube. He remarked on Győr's continued commercial importance.[105] The city certainly benefited from several new industries founded between the 1860s and the 1910s, such as the cookie biscuit factory that eventually employed 1,000 people before World War I, food processing facilities for oil and vinegar founded in the mid-1880s, plants producing farm equipment, and a sugar processing plant employing nearly 500 people, amid many others—39 in total.[106] Baross also noted that the city's steam navigation would benefit from public warehouses on the river, a project that materialized in the late 1890s.[107]

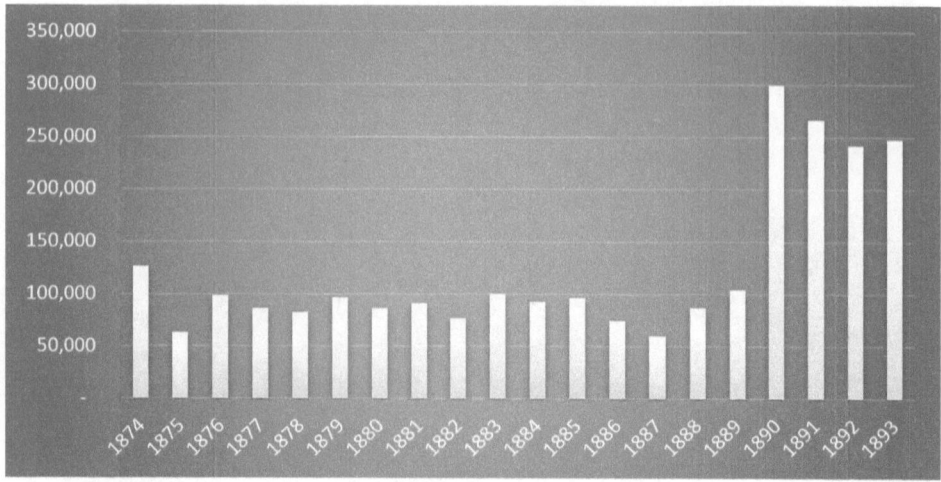

Figure 5.9. Győr Steam Navigation Company Freight Transportation (in tons), 1874–1892. *Source:* Annual statistical reports (*Magyar Statistikai Évkönyv*) issued by the Hungarian Statistics Bureau since 1870.

The Győr Steam Navigation Company's prospects rose with the transshipment hub, the proliferation of industries, and the local regulation works. Although shareholders complained about conditions in their local branch, the Small Danube, during their annual meeting in 1891, arguing that large-scale freight shipping was not possible, the company also reported a profit of more than thirteen thousand florins for the year.[108] An injection of capital and expansion of ship ownership also expanded the company and its operations quite significantly in the early 1890s. The boost to the Győr Steam Navigation Company's business prompted the city council to likewise approve warehouse construction in 1893 (fig. 5.9).

The company's prosperity and rising prospects boded well for the community until national politics intervened the following year. In 1894, the government in Budapest founded a national steam navigation company, the Hungarian Royal River and Sea Navigation Joint-Stock Company (MFTR), which set about consolidating Hungary's plethora of small regional companies. Seeking to tap into Győr's steady traffic and lively business, the MFTR bought the Győr Steam Navigation Company, its ships, and its shipyard. The end of Győr's independent steamboat company did not halt the city's momentum in building up its Danube infrastructure. The opening of public warehouses in 1899 contributed to the revival of commerce. Steamboat companies

Act Locally, Think Imperially

Figure 5.10. Győr's Mosoni Danube, the Landing Place (center right), and the Grain Elevator (far right). *Source: Üdvözlet Győrből* [Greetings from Győr], early 1910s. Postcard. © Dr. Pál Kovács Library, Győr, Hungary.

from Germany, Austria, and Hungary celebrated the new facilities. In a nod to the city's growing fortunes, the city also constructed a modern grain elevator in March 1914 (fig. 5.10).

As commercial relations improved in cities along the Danube, the opportunity for greater recreational pursuits along the river followed. Linz's regulation of the Danube in the last decades of the century reclaimed new land that became available for public use. The Swim and Bathing Facilities settled near the riverside Elisabeth Quay, raising over 200,000 crowns to build separate spaces for men, women, and military cohorts. Another river bathhouse upstream expanded its facilities in 1907, and hundreds of students visited it for swim school.[109] The city built a new promenade along the Danube's regulated banks where the Beautification Club was responsible for planting trees and setting up benches for walkers. Likewise, next to the new steamboat landing site, the city created a large park for its residents. In 1892, the new club Youth Games and Physical Fitness (*Jugendspiele und Körperpflege*) regularly used the immense space to encourage children to get out and exercise, through gymnastics, rowing, and swimming, and in the winter ice skating and pulling sleds on the river. Eventually, in 1909, the city even donated a large area along the Danube to the club.¹¹⁰

In Győr, the public had admittedly appreciated recreational opportunities on the Danube for decades. In 1869, nearly three hundred men in town formed the Győr Boat Club (Győri hajós egylet) and in 1877, one hundred members started the Boating Association. In the 1880s, Győr's boating clubs participated in regattas that met up and down the Danube, competing with other towns in the empire's Austrian and Hungarian halves.¹¹¹ Its local associations endeavored to organize yearly regattas as well for yachts and rowing crews, and even novelty events, like racing "boats carved from tree logs."¹¹² Extraordinarily, the rowing club served not only a recreational but also a civic purpose; during times of flooding, members of the club took out their boats to aid rescue efforts. When younger members of a youth boating club took to the waters in 1897, the *Pesti Hírlap* declared that elementary schools in Pozsony/Pressburg should encourage the formation of clubs based on "Győr's model."¹¹³

Along the river's banks, members of the public also enjoyed all manner of activities from swimming to biking.¹¹⁴ One correspondent joyfully described the scene at the turn of the century:

> The Danube, Rába, Rábca and the Marcal flow into city. . . . The rapid Rába river is very dirty and cloudy, and there are no bathhouses on it. Free swimmers who do not like the foam of the blond Danube . . . bathe in it. Here is the Rábca which is no ordinary river but contains iodine in its medicinal brown flow. During the bathing season, old gentlemen with gout and plebians without gout take a dip in it. The specialty of Győr's clerks and prize-winners is the Balaton Bath, near the Hops Garden Restaurant . . . these are all specialties of Győr.¹¹⁵

With all these chances to swim in open segments of the river, it nevertheless rankled students that they lacked a permanent swimming facility to practice for their tournaments. Representatives from the youth rowing club petitioned the national authorities for funds to build a floating bathhouse but their request was declined. When flooding and ice floes destroyed the rowing club's facilities, the students considered an alternate arrangement. A local engineer suggested that they build a facility on the river's banks rather than one floating in the river. Armed with the plans that he provided to them, the club's members approached the city for permission to build at a new location on the Rába River. The council did not outright reject their request but explained that the local Rába Regulation Company held a permit for building along the proposed stretch. The company refused to grant the students permission to build there, dashing their hopes.[116] A brand new swimming facility was eventually opened a few decades later in the interwar period.

The meandering circumstances that residents in Linz and Győr experienced throughout the nineteenth and early twentieth centuries were representative of the many challenges and opportunities that municipalities in the Habsburg Empire felt. Hoping to promote trade and industry, businesses proposed engineering works that prioritized the river's use in transportation and industrial processes. Engineers, civil servants, certain businesses, and members of the public, on the other hand, recognized the need to protect the populace from flooding and disease epidemics. Such protection also entailed the acculturation of riparian residents, so they conducted themselves in a way that benefited the general well-being. Through all this, people also increasingly took to the river for recreation and escape, enjoying the newly created spaces, activities, and services that companies, municipal authorities, and associations offered. In all these instances, local communities grappled with a river that sometimes acted independently of their desires and plans. To safeguard themselves against this unpredictability, citizens sought out partners in local guilds, companies, associations, and city councils as well as with transnational, commercial, and imperial bodies that could help protect their interests on the river. While each city's unique history shaped its development along the Danube, the river served as a common feature for those people, a source of shared experiences for municipal populations throughout the empire.

Economic and Civic Engagement Along the Danube

In the early 1880s, Habsburg Crown Prince Rudolf initiated an encyclopedic series titled *The Austro-Hungarian Monarchy in Word and Image* as a means of educating inhabitants of the empire about the diversity and history of the realm.[117] Besides the project's academic intent, Crown Prince Rudolf also envisioned that descriptions of the empire would provide "for the elevation of universal, patriotic love" to encourage "feelings of solidarity which would bind together all the people of the fatherland."[118] In the first of what would eventually total twenty-four volumes, several prominent scholars wrote about the Danube's regulation and its influence on Vienna's development:

> The time is too short, to be able to judge the great worth that the Danube's regulation signifies for trade and travel [*Handel und Wandel*] at the capital.... Trade and industry are slowly transferring to the banks of the mighty river, in the newly developed *Donaustadt*. Many streets lead to this previously undeveloped part of the capital, rail lines are starting to connect the banks of the river with the train network of the empire, industrial establishments are starting to develop there to process raw materials and mass goods, [and] storehouses have already been erected... to be filled with mass goods needed for the Viennese wholesale trade. New life is stirring on the banks of the majestic river, which is no longer threatening in its new bed, and instead only more fruitfully and usefully serves to cultivate the welfare of the whole empire. Out of the restlessly busy drive, which is awakening in the *Donaustadt*, a glorious Vienna on the magnificent Danube is blooming.[119]

This description most certainly exaggerated certain aspects of this industrial and commercial growth, but the enthusiasm matched the optimism that many residents of the city felt about these massive new projects taking place and reshaping the imperial capital. Residents looked at the river's regulation as a new source of land on which to build more housing while business leaders envisioned factories and commercial hubs. Many probably hoped that reclaimed land would protect against floods or looked at the river's regulation as a solution for unsanitary urban conditions. Most expected to enjoy the river for leisure and amusement.

And they were not alone. For many cities on the Danube, like Linz and Győr, engineering works combined with new technical innovations, like steam-powered boats, excavators, and cranes, to create all new spaces on the river from bridges to quays to commercial hubs to swimming facilities. Local projects on the Danube increasingly took on greater significance, as residents looked to provincial, national, and imperial authorities to approve and financially support their initiatives. Because of the fundamental effect that such works had on people's lives and livelihoods, local and imperial authorities also played a critical role mediating disputes and establishing fair access to and usage of the river for many different stakeholders. Commercial interests, political agendas, and concerns of the public were all part of the negotiations about the manner and extent of these engineering works and urban renewal projects on the Danube. What these discussions revealed were two critical developments for citizens of the empire. First, the presence of functional civic engagement between governments and their people heard concerns and, where possible, addressed them. Second, the robust exercises of a pluralistic society contained divergent ideas and different people, who nevertheless strove for compromise and the "common good" for citizens and civil society along the Danube.

Conclusion

COLLECTIVE ACTION AND THE COMMON GOOD

At one o'clock in the morning on July 29, 1914, two Austro-Hungarian river monitors released the opening salvos of World War I when they bombarded fortifications at Belgrade from the Danube River. Given the Danube's centrality to the Habsburg Empire's history, it is perhaps fitting that the river was the opening scene of a conflict that would eventually end the empire. Historians are quick to caution, however, that just because the empire collapsed in 1918, the events and history preceding it did not ineluctably lead to its decline, nor did its dismemberment erase its social, economic, and political legacy.[1] On the contrary, countless newspaper articles, commercial guides, associational presentations, parliamentary debates, technical reports, and celebrations indicated a strong and enduring connection among the disparate peoples and regions of the Danube Empire.

Outside observers noticed this connection as well. When World War I ended, the British Admiral Ernst Troubridge—the future head of the Inter-Allied Danube Commission—recommended that food

Conclusion

and supplies begin flowing as soon as possible along the Danube to relieve the war-weary population. As a result, the Supreme Economic Council, which was formed to stabilize the postwar economic situation in Europe, declared, "Despite the existing uncertainty concerning the frontiers and the ownership of the floating material, normal conditions of traffic on the Danube should be re-established as soon as possible."[2]

The work done by citizens, businesses, and governments forged this enduring legacy over the nineteenth and early twentieth centuries. As this book has explored, it was not always straightforward: the empire's stability was tested by national and liberal ideologies and by growing social discontent among the working classes. The Danube, wild and mostly unregulated for half of the century, visited disasters and catastrophes upon its numerous riparian communities. The Habsburg regime countered these natural and political threats with reforms and improvements to shore up support for and reaffirm the empire's legitimacy. Newfangled steam power and large-scale hydraulic engineering projects were employed to make the river safer and more useful to the inhabitants of the empire.

Certain changes to the river had negative consequences—both socially and environmentally. Mark Cioc says in *The Rhine: An Eco-Biography* that "too often economic historians see only the steamship's forward progress, while environmental historians see but the smokestack's plume. Depending on perspective, a fishing village can be depicted as pleasingly arcadian or hopelessly backward. . . . Such simple abstractions, however, mask more than they reveal: rarely are the lines separating progress and decline . . . so easily drawn."[3] Human activity, from regulation works to agricultural practices, affected the river's hydrology with scant concern for the long-term consequences. New practices on the river, like steam navigation drove traditional boat operators out of business and disrupted fishing and milling operations along the river.[4] Rivalries between the Danube Steam Navigation Company (DDSG) and its Hungarian competition led at times to acrimonious debates about state-sponsorship and fairness.[5]

The Danube retained its prominent agency in this history. Even when the river's regulation was near completion by the end of the nineteenth century, climatic variability continued to flood and diminish river levels, while cold snaps and ice floes led to unavoidable disruptions.[6] As Peter Coates cogently observes in *A Story of Six Rivers*, we should not be surprised that rivers can have agency, as we frequently personify them—calling them lazy, violent, proud, dangerous—and ascribe to them characteristics and powers that we do not normally afford other nonhuman actants.[7] Rivers move objects as large as ships and as small as sand particles, they wear away at their banks, beds, and any structures within their path, and they flow, sink, and flood. Their movement and activity are not passive, but they prompt reactions and guide behavior. With eighty or so cities and communities connected by steamboat station along the Danube, and hundreds of other cities on tributaries across the empire, from Arad on the Maros River arising in eastern Hungary to Kufstein on the Inn River in Tirol in the west, millions of people spanning the empire were subject to the river's behavior.[8]

People were, unsurprisingly, also active participants in the river's transformation. The river was both a physical and a socioeconomic space where the interests and needs of the authorities and the empire's population intersected. Discussions regarding hydraulic projects also involved active input from commercial, associational, and civil groups. Petitions, newspaper articles, associational activity, speeches, and interactions with the governments indicated that citizens not only *supported* certain interventions along the river but also *expected* them. When stakeholders' interests did not align, for professional or technical or economic reasons, governmental mechanisms and institutions stepped in to navigate these conflicts, in the process determining how well they retained the trust and loyalty of the people. Citizens' actions reveal that despite barbed political exchanges or frustrations that emerged, people relied on and operated within the structures of the empire to pursue their interests, which ultimately served as a daily plebiscite reaffirming the Danube Empire's legitimacy to exist.[9]

Today, the force of our political debates regarding the environment have taken on a different tone and have grown on a much grander scale. In *Something New Under the Sun*, John McNeill points out that while economic output, population growth, and energy usage have grown consistently for hundreds, if not tens of thousands of years, they spiked so drastically in the twentieth century that we radically altered our relationship with the earth.[10] The disproportionate effect humans have had on the environment has led many to designate this new era "the Anthropocene" in recognition that the impact of humans on the earth's biogeophysical processes has overtaken that of nature.[11] This has grave consequences beyond the pollution and despoliation of our earth's ecosystems. The accumulation of excess greenhouse gases such as carbon dioxide, methane, and nitrous oxide—by-products of our agricultural, commercial, and industrial activities—have changed the chemical composition of our atmosphere to such a degree and with such long-lasting effect that the scientific community and large swaths of the global society agree that we face an existential threat of our own making. With anthropogenic gases in our atmosphere trapping and retaining enough solar energy to raise mean global temperatures, we are on the precipice of tipping the atmosphere into runaway greenhouse gas accumulation. With the proverbial ship already in midstream, however, it is already no longer a question whether flooding or other natural disasters will occur but how frequently and how intensely they will hit.

In a way, it is ironic that we are in this position. So much of human activity in earlier centuries was geared toward modifying the environment and taming nature to guarantee our survival, secure energy needs, and advance our economic, cultural, and political development. And yet those modifications and the mentality behind them have led directly to our current vulnerabilities. For years, western US states have faced widespread drought and wildfires as the large river and irrigation networks developed for decades to supply water to some of the driest parts of the country have faltered and failed in the face of dwindling rainfall and water levels. In China, heavily engineered rivers that were dammed to prevent floods and provide hydropower are nevertheless flooding and endangering millions of lives as heavier rainfall threatens these constructs.[12] Furthermore, in much the same way that riparian communities failed to overcome their provincial concerns or set aside local interests to collectively fight against Danube flooding, climate change today represents an existential threat for countries offering piecemeal efforts that fail to grasp the urgency or address the need for holistic changes in our relationship with the environment. Current debates about fighting climate change are multifaceted, straining multilateral diplomacy and fueling political partisanship. For activists, it feels like a battle that pits citizens against corporations and (at times) the government. Developing countries place the responsibility for addressing climate change at the feet of industrialized nations, which have been responsible for many of the historical emissions that have led to this crisis.[13] Some of the largest polluters—

Conclusion

China and the US—continue to rely heavily on coal, natural gas, and oil, despite (sporadic) promises to reduce fossil fuels and policies to promote shifts to green energy.[14]

However, just as citizens and rulers alike found common ground on the Danube, so too our climate crisis can represent a crucial opportunity for humans. The Danube's complexity certainly represented a political problem as much as it did a technical-environmental one. And yet stakeholders eventually came to realize that without collective action, civic engagement, and transnational cooperation, it would not be possible to address challenges or benefit from the ecological, energy-oriented, and material conditions that the river provided. Now the question is: What will climate change mean for our current generation? Can it herald an age of more sustainable energy production, greater global cooperation, and improved well-being for all? Perhaps it can, if citizens realize the value of collective action, hold governments accountable and urge them to implement policies that support the "general well-being" of all, and work together to construct a more just and sustainable global community.

NOTES

Introduction: The Danube Empire

1. Anonymous, "Zur Gründung des Donau-Vereins," *Neue Freie Presse*, March 25, 1879, p. 9.

2. Anonymous, "Das Zweck-Essen des Donauvereins," *Die Presse*, June 10, 1879, p. 9. Translations in this book are mine unless otherwise indicated.

3. Gary B. Cohen iterates, however, that nationalism did not threaten the integrity of the empire, arguing "for their part, the nationalist and other popular political forces in the Habsburg Empire during the late nineteenth century were hardly irresistible forces demanding a self-government that could be realized only by dissolving the empire. Indeed, the great majority of the political parties and organizations up until World War I contended for greater empowerment within a reformed multinational Habsburg state, not for independence." Cohen, "Nationalist Politics," 242.

4. Anonymous, "Das Zweck-Essen des Donauvereins."

5. Anonymous, "Das Zweck-Essen des Donauvereins." The Habsburg Empire (named for the Habsburg monarch rulers) was also called the Austro-Hungarian Empire after the 1867 Compromise split internal governance of the empire with capitals at Vienna and Budapest. The terms for the two halves of the empire "Austria" and "Hungary" are not coterminous with the modern countries, as "Austria" included the Alpine lands roughly corresponding to today's Austria, along with the Bohemian lands, Dalmatia (the modern Croatian coastline on the Adriatic), and Galicia (Austria's shared of partitioned Poland), while "Hungary" also included Croatia-Slavonia with

its southern Slavs, Transylvania with its amalgamation of different ethnicities, and large Serb communities in the Banat and Vojvodina. The border river dividing these two halves of the empire, the Leitha, gave rise to the monikers "Cisleithania" "Transleithania" for Austria and Hungary, respectively.

6. Anonymous, "Zur Donauregulierung," *Wiener Landwirtschaftliche Zeitung*, July 5, 1879, p. 278.

7. Anonymous, "Hirek," *Ellenőr*, September 10, 1879.

8. Anonymous, "Hirek," *Ellenőr*, August 14, 1879.

9. Anonoymous, "Kell-e Dunaegylet," *Pesti Hírlap*, October 27, 1879, pp. 2–3.

10. János Hunfalvy, "Zur Frage der Flussregulierungen II," *Pester Lloyd*, November 30, 1879.

11. Gingrich et al., "Danube and Vienna"; and Hohensinner, Drescher et al., *Genug Holz für Stadt und Fluss?*

12. Rácz, *Steppe to Europe*, 218–30.

13. Hohensinner and Hauer, "Durchstich, Kai und Häusergerümpel."

14. Veichtlbauer, "Port of Vienna"; and Gierlinger et al., "Feeding and Cleaning the City."

15. Hagen and Hauer, "Hygiene und Wasser."

16. Exner et al., "Volkswirthschaftliches Leben in Wien," 1:321.

17. A. W. Mitchell, *Grand Strategy*, 26–33.

18. Blanning, *Joseph II*, 76–79; and Ingrao, *Habsburg Monarchy*, 212–15.

19. Veres, "Constructing Imperial Spaces," 63–64.

20. Jászi, *Dissolution of the Habsburg Monarchy*, 189–90.

21. Tschischka, *Der Gefährte auf Reisen*, vi.

22. Müller, *Die Donau vom Ursprunge*, vii.

23. Wilson, "Lost Opportunities."

24. Czoernig, *Österreich's Neugestaltung*, 321–22.

25. Dürrnberger, "Das Donauthal von Passau bis Linz," 6:10.

26. Andrásfalvy, *A Sárköz és a környező Duna*. Andrásfalvy concludes that it was only with systemic embanking in the mid- to late nineteenth century that these dangers abated.

27. Glassl, "Der Ausbau der ungarischen Wasserstraßen," 46.

28. Gruffat, "Beautiful Public Danube."

29. Borsos, "Rivers, Marshes, and Farmlands."

30. Jungwirth et al., *Österreichs Donau*, 185–212; and Csukovits and Forró, *Duna*.

31. George Ritter von Frauenfeld, "Die Wirbelthierfauna Niederösterreichs," *Jagd-Zeitung*, August 31, 1871.

32. Richard Genthner, "Stillleben am Donauufer," *Neues Wiener Journal*, April 3, 1923.

33. Földes, *Nemzetgazdasági Statisztikai Évkönyv*, 102; and K.k. Statistische Central-Commission, *Statistische Monatschrift: XII*, 309.

34. *Die Donau und ihr Einzugsgebiet*, 47–50, 166–67.

35. Andrásfalvy, "Die traditionelle Bewirtschaftung," 39.

36. Rácz, "Climate History of Central Europe," 242; Hohensinner and Haidvogl, "Zu viel Wasser," 161; and McCarney-Castle et al., "Simulating Fluvial Fluxes," 94.

37. Giosan et al., "Early Anthropogenic Transformation," 5.

38. Hohensinner and Schmid, "The More Dikes, the Higher the Floods."

39. Kornél Szokolay, "A pozsony gönyöi Dunárész szabályozása," *Pesti Hírlap*, February 23, 1880, p. 6.

40. Coen, *Climate in Motion*.

41. "Volkswirthschaft," *Wiener Presse* (Vienna, Austria), January 6, 1884.

42. Gonda, *Az al-dunai Vaskapu*, 111–12.

43. Braudel, *Civilization and Capitalism*, 70–92.

44. McNeill, *Something New*, 10–11.

45. Rhodes, *Energy*, 25–33.

46. Malm, *Fossil Capital*, 18.

47. McNeill and McNeill, *Human Web*, 236–48.

48. Ross, *Power and Ecology*.

49. Komlosy, "Innere Peripherien."

50. Malm, *Fossil Capital*, 3–7.

51. Deák, *Beyond Nationalism*; Shedel, "Emperor, Church, and People"; King, *Budweisers into Germans and Czechs*; Judson, *Guardians of the Nation*; and Zahra, *Kidnapped Souls*.

52. Judson, *Habsburg Empire*; Deak, *Forging a Multinational State*; Shedel, "Mother of It All"; and Unowsky, *Pomp and Politics*.

53. White, *Organic Machine*.

54. Worster, *Rivers of Empire*; and Blackbourn, *Conquest of Nature*.

55. Husain, *Rivers of the Sultan*.

56. Pritchard, *Confluence*.

57. Hohensinner, Herrnegger et al., "Type-Specific Reference Conditions"; Hohensinner, Jungwirth et al., "Spatio-Temporal Habitat Dynamics"; Gingrich et al., " Danube and Vienna"; Hohensinner, Sonnlechner et al., "Two Steps Back"; Jungwirth et al., *Österreichs Donau*; and Haidvogl et al., *Wasser, Stadt, Wien*.

58. Renan, "What Is a Nation?"

Chapter 1. Creating the Imperial Danube

1. Anonymous, "Vermischte Nachrichten," *Österreichischer Beobachter,* October 22, 1823.

2. Good, *Economic Rise,* 58.

3. Anonymous, "Wien, den 11. October," *Österreichischer Beobachter,* October 12, 1823.

4. Grössing et al., *Rot-Weiss-Rot,* 9.

5. Anonymous, "Wien, den 6. November," *Österreichischer Beobachter,* November 6, 1823, 1408–10.

6. Uekoetter, *Turning Points of Environmental History,* 2.

7. Koselleck, "Einleitung."

8. Wagar, *Double Emperor,* 118–21.

9. Edward Timms discusses the way that the Enlightenment created the notion of "public opinion" and how Metternich was particularly adept at employing it to legitimate Habsburg authority and the "Austrian" idea. Timms, "National Memory," 901.

10. Ingrao, *Habsburg Monarchy,* 152–92.

11. Katalin, "Later Ottoman Period," 102.

12. Blackbourn, "'Time Is a Violent Current,'" 13.

13. Blackbourn, *Conquest of Nature,* 2–75.

14. John Rae, *Life of Adam Smith,* quoted in Viner, "Adam Smith and Laissez Faire," 200.

15. *Huse v. Glover,* 119 U.S. 543 (1886).

16. Haselsteiner, "Cooperation and Confrontation."

17. Komlos, *Habsburg Monarchy,* 4.

18. Kállay, "Ungarischer Donauhandel," 45.

19. Hohensinner, Lager et al., "Changes in Water and Land," 159.

20. Glassl, "Der Ausbau der ungarischen Wasserstraßen," 47.

21. Quoted in Kállay, "Ungarischer Donauhandel," 44.

22. Jungwirth et al., *Österreichs Donau,* 185–212.

23. Jokai, *Timar's Two Worlds,* 11.

24. *Edinburgh Encyclopedia,* s.v. "Danube" 575.

25. Zerlik, "P. Joseph Walcher, S.J."

26. Walcher, *Nachrichten von den bis auf das Jahr 1791,* 22.

27. Ricci, *China in the Sixteenth Century,* 12.

28. Voltaire, *History of Peter the Great,* 334.

29. Mann, *Treatise on Rivers,* 4.

30. In his work, Martin Doyle also details how in the late 1780s, George Washington, Thomas Jefferson, and others in the young American republic were surveying western territories to discover a way to unite the Ohio River and the country's fertile interior with coastal commerce, hoping to dispel North–South rivalries and overcome European trade competition. Doyle, *Source*, 17–34.

31. Blackbourn, *Conquest of Nature*, 89.

32. Judson, *Habsburg Empire*, 28–29.

33. Veres, "Constructing Imperial Spaces," 30.

34. Veichtlbauer, "Von der Strombaukunst zur Stauseenkette."

35. The Iron Gates was a segment of the Danube where rocky cataracts made navigation extremely difficult. Hajnal, *Danube*, 99–103.

36. Mauch, "Introduction"; and Dénes, *A magyar vízszabályozás*, 78.

37. Franz's brother Joseph, the Hungarian palatine, intervened in 1810 to ensure the establishment of the Sárvízi Csatorna Társulat (Sárvízi Canal Company), which had more success raising funds to undertake the Sárvíz regulation. Fejér, *Árvizek és belvizek szorításában*, 17.

38. Dóka, "A vízügyi szakigazgatás fejlődése, I," 518.

39. Glassl, "Der Ausbau der ungarischen Wasserstraßen," 64.

40. The canal's location passes through modern-day Serbia in the semi-autonomous northern province of Vojvodina. The Danube and Tisza were the two longest rivers in the empire.

41. Beszédes, *Duna-Tiszai hajózható csatornáról*, 4–6.

42. Vedres, *Ueber einen neuen Schiffbaren Kanal*, 3.

43. Gonda, *Die ungarische Schiffahrt*, 13.

44. Dénes, *A magyar vízszabályozás*, 72.

45. Fray, *Allgemeiner Handlungs-Gremial-Almanach*, 488.

46. Oelwein, *Die Binnen-Wasserstrassen*, 3.

47. Czeike, "Wiener Neustädter Kanal," 639.

48. Hauer et al., "How Water and Its Use Shaped," 317.

49. Czeike, "Wiener Neustädter Kanal," 639; and Hauer et al., "How Water and Its Use Shaped," 319.

50. Fray, *Allgemeiner Handlungs-Gremial-Almanach*, 327.

51. Good, *Economic Rise*, 34.

52. Grössing et al., *Rot-Weiss-Rot*, 11.

53. Grössing et al., *Rot-Weiss-Rot*, 12.

54. Good, *Economic Rise*, 64.

55. Sisa, "Count Ferenc Széchényi's Visit."

56. Széchenyi, *Über die Donauschifffahrt*, 15.

57. Csikvári, *A közlekedési eszközök*, 120.

58. Dezsényi and Hernády, *A magyar hajózás*, 57; and Hajnal, *Danube*, 134.

59. Hajnal, *Danube*, 128.

60. Széchenyi, *Über die Donauschifffahrt*, 75.

61. Trautsamwieser, *Weisse Schiffe*, 15–20.

62. Günther Harum, "Das war unsere Donau," unknown newspaper clipping, September 21, 1950.

63. K.k. Direction der administrativen Statistik, "Dampfschiffahrt und Eisenbahnen," *Tafeln zur Statistik . . . Jahr 1842* (1846), 445–53. Note that "k.k." stands for *kaiserlich-königlich* (imperial-royal) and refers to the emperor's dual authority as emperor of the Austrian Empire and the king of specific crownlands, including Bohemia, Moravia, Silesia, and Hungary.

64. Hajnal, *Danube*, 62, 145.

65. Pisecky, "Die Grösste Binnenreederei," 55–56.

66. Erste k.k. priv. Donau-Dampfschiffahrts-Gesellschaft, *Geschäfts-Bericht . . . 1878 bis . . . 1879* (1880), 3.

67. This rough estimation is according to the currency conversion algorithm of Rodney Edvinsson, an associate professor at Stockholm University, and it compares various conversion rates of florins into USD from 2015 then adjusted for inflation in 2024. Edvinsson "Historical Currency Converter." Dezsényi and Hernády, *A magyar hajózás*, 99.

68. DDSG Administration to the Imperial-Royal Trade Minister, December 3, 1890, AT-OeStAAVA Inneres MdI Allgemein A 459.

69. Jekelfalussy and Vargha, *Közgazdasági és statisztikai évkönyv*, 431–44, 450.

70. Based on data published in the Hungarian Statistical Office's yearly *Statisztikai évkönyv*, which collected passenger and freight data for registered steam navigation companies operating in the Kingdom of Hungary. According to an aggregation of the data, while the Hungarian Royal River and Sea Joint-Stock Company (MFTR) transported 445,544 passengers between 1895 and 1914, the DDSG transported 510,600 passengers.

71. Unowsky, *Pomp and Politics*, 1–10.

72. Anonymous, "Különfélék," *Pécsi Figyelő* (Pécs, Hungary), January 3, 1880.

73. Sartori, *Wien's Tage der Gefahr*.

74. "Wien," *Wiener Zeitung* (Vienna), March 5, 1830.

75. Anonymous, *Die Ueberschwemmung zu Pesth*, 37.

76. This approximation takes the average from Bohemia, which was the middle wage bracket between the higher Viennese wages and the wages in Galicia, which could be 50 percent lower. Jobst and Stix, "Florin, Crown, Schilling," 105–6.

77. Unowsky, *Pomp and Politics*, 116.

78. Ziegler, *Gallerie*.

79. "Der Kaiser auf dem Marchfeld," *Tages-Post*, June 11, 1905.

80. Hakkarainen, "City Upside Down."

81. Herold, *Brigittenau*, 33.

82. "Theresa Krones: Zu ihrem hundertsten Geburtstag," *Welt Blatt* (Vienna), October 8, 1901.

83. Fassmann, "City-Size Distribution," 11.

84. Bousfuield and Toffoli, *Royal Tours*, 42–56.

85. Blackbourn, *Conquest of Nature*, 189–249.

86. Grever, "Staging Modern Monarchs"; Stephanov "Ruler Visibility"; Blackbourn, *Conquest of Nature*; and Truesdell, *Spectacular Politics*, 81–100.

87. Hamann, "Empress Elisabeth," 135.

88. Anonymous, "Nicht Amtlicher Teil," *Wiener Zeitung* (Vienna), April 23, 1854.

89. Beller, *Francis Joseph*.

90. Anonymous, *Wien seit 60 Jahren*, 17.

91. Anonymous, "Nichtamtlicher Theil. Kärnten," *Klagenfurter Zeitung*, September 5, 1856.

92. "Communal Zeitung, Inaugurirung der Donau-Regulirungs-Arbeiten," *Neue Freie Presse*, May 14, 1870.

93. "Feierliche Eröffnung der Schiffahrt im neuen Strombette der Donau," *Wiener Zeitung*, May 31, 1875.

94. C. Szabó, "Brücken über die Donau," 105.

95. Friedrich Paul, "Die Augarten-Brücke über den Wiener Donaukanal," *Allgemeine Bauzeitung*, (Vienna), 1881.

96. Exner et al., "Volkswirthschaftliches Leben in Wien," 1:322.

97. In 2015 prices, 12.5 million florins was equivalent to about 170 million USD (Edvinsson, "Historical Currency Converter"); Lukács, "Közlekedési intézmények," 514.

98. In 1890, the DDSG recorded its highest level of traffic at over 3.5 million passengers, and the Hungarian Statistical Office recorded tens of thousands of annual passengers for the DDSG and Hungarian River and Sea Navigation Company on tributaries like the Drava and Tisza.

99. The *honfolalás* or "Homeland Conquest" referred to the Magyars' ar-

rival in the Carpathian Basin in approximately 896, though a committee arbitrarily decided the year to celebrate. The Franz Joseph Bridge was the city's third bridge to span the Danube. The Chain Bridge, Budapest's first permanent bridge, had opened in November 1849 and the Margít Bridge opened in 1876, but both remained woefully insufficient for city traffic, with ferry services transporting millions of people across the Danube between Buda and Pest each year. As part of Buda and Pest's revitalization, Hungary's Law X (törvénycikk X) in 1870 mandated more bridges on the Danube.

100. "A Ferenc-József híd," *Pesti Napló*, October 2, 1896; and "Einweihung der Franz-Joseph Brücke," *Pester Lloyd*, October 5, 1896.

101. *Vasárnapi Újság*, October 11, 1896.

102. "Se. Majestät der Kaiser in Pressburg," *Wiener Zeitung*, December 30, 1890.

103. "Hidavatás Pozsonybany," *Budapesti Hírlap*, December 30, 1890.

104. Komárom is today's Komárno, Slovakia, and Új-Szőny is modern-day Komárom, Hungary.

105. Gonda, "A magyar duna," 35.

106. "A Mária Valéria híd története," *Felvidék.ma*, last modified November 25, 2013, http://felvidek.ma/2013/11/a-maria-valeria-hid-tortenete/; and Mutschlechner, "Franz Joseph."

107. Laurencic, "In Linz und Gastein," 558–59.

108. Anonymous, *Jubiläums-Ausstellung in Wien*.

109. Csorba, "Transition," 107.

110. The *Vasárnapi Ujság*'s nineteenth paper in 1898 had a large image of the projected bridge along with an article about its construction. For the story of Franz Joseph's presence at the foundation laying, see Szabó, "Brücken über die Donau," 98.

111. "Napi hírek," *Budapesti Hírlap* (Budapest), December 18, 1898.

112. The *Budapesti Hírlap* mentioned the opening in its papers September 20, 25, 29, 1903.

113. For example, in Komárom, along the Danube's bank next to Elisabeth Bridge, a nice, green space became Elisabeth Park. In Göndöllő, one of Elisabeth's favorite retreats, Franz Joseph opened Elisabeth Park in 1901. Bałus, *Stadtparks*, 154.

114. *Rechenschaftsbericht des Gemeinderathes*, 305.

115. "Erzsébet királyné szobra," BudapestCity.org, https://budapestcity.org/egyeb/tervek/Erzsebet-kiralyne-szobra/index.html.

116. Shedel, "Mother of It All."

Chapter 2. The Danube as Life Artery

1. Zeilinger, "Zur Urheberschaft der Pasetti-Karte."
2. Klun, "Flusskarten der Donau," 2.
3. Klun, *Allgemeine Geographie*; and "Klun, Vinzenz," 425.
4. Coen, *Climate in Motion*, 63–118.
5. Klun, "Flusskarten der Donau," 17.
6. *Reise auf der Donau*, 23.
7. *Reise auf der Donau*, 70.
8. Hohensinner, "'Sobald jedoch der Strom,'" 42.
9. Malte-Brun, *Universal Geography*, 161.
10. Ágoston, *A nemzet inzsellérei*, 12.
11. Dóka, *A vízimunkálatok irányítása*, 114.
12. Szabadfalvi, "Nomadic Wintering System."
13. Biró et al., "Reviewing Historical Traditional Knowledge," 1121.
14. Borsos, "Rivers, Marshes, and Farmlands," 199.
15. Malte-Brun, *Universal Geography*, 160.
16. Judson, *Habsburg Empire*, 228.
17. Prechtl, "Zur Geschichte der Dampfboote," 214.
18. In 1829 total expenditures was 1,631,118 florins; in 1830 it was 1,875,999; and in 1831 it was 2,137,378. "Strassen- und Wasserbau Aufwand," *Tafeln zur Statistik der österreichischen Monarchie, 3. Jahrgang 1831* (Vienna); and Jobst and Stix, "Florin, Crown, Schilling," 117.
19. Széchenyi, *Über die Donauschifffahrt*, 5–30.
20. Fray, *Allgemeiner Handlungs-Gremial-Almanach*, 327.
21. "Ungarn," *Pressburger Zeitung* (Bratislava), January 21, 1840; and "Magyarország," *Hazai 's Külföldi Tudósítások—Nemzeti Ujság* (Budapest), January 25, 1840.
22. Pach, "Széchenyi és az Alduna-szabályozás," 559.
23. Mészáros, *Gróf Széchenyi István*.
24. DDSG Circular, November 18, 1833, Széchenyi iratok 28-17.190, Magyar Környezetvédelmi és Vízügyi Múzeum [Hungarian Environmental Protection and Hydrological Museum], Esztergom, Hungary.
25. Pach, "Széchenyi és az Alduna-szabályozás," 570.
26. Baader, *Ueber die Verbindung*; Soden, *Der Maximilians-Canal*; and Kleinschrod, *Die Kanal-Verbindung*.
27. Nemes, *Another Hungary*, 81–86.
28. Lower Austria, Upper Austria, Styria, Carinthia and Carniola, the

Austrian Littoral, Tirol, Bohemia, Moravia and Silesia, Galicia, and Dalmatia. The *Tafel zur Statistik* did not include Hungary either.

29. Veichtlbauer, "Von der Strombaukunst zur Stauseenkette," 64.

30. In 1842, these two figures alone made up more than 20 percent of all "additional ship costs" outside the ships' maintenance expenses, and by 1843, this had risen to almost 25 percent of additional costs, but dropped to only 14 percent in 1844 (the expenses were similar, but *other* "additional costs" increased), K.k. Direction der administrativen Statistik, *Tafeln zur Statistik ... Jahr 1842*; K.k. Direction der administrativen Statistik, *Tafeln zur Statistik ... Jahr 1843* (k.k. Hof- und Staatsdruckerei, 1847); and K.k. Direction der administrativen Statistik, *Tafeln zur Statistik... Jahr 1844*.

31. Dorner, *Panorama der Österreichischen Monarchie*, 11.

32. Paget, *Hungary and Transylvania*, 187–88.

33. Hohensinner, "'Sobald jedoch der Strom,'" 53; and Veichtlbauer, "Von der Strombaukunst zur Stauseenkette," 69.

34. Dóka, *A vízimunkálatok irányítása*, 71.

35. Barany, "Age of Royal Absolutism," 203.

36. Dóka, *A vízimunkálatok irányítása*, 71.

37. Rácz, "Danube Pontoon Bridge of Pest-Buda," 472–81.

38. István Széchenyi to Palatine Joseph, May 29, 1847, Széchenyi iratok, Magyar Környezetvédelmi és Vízügyi Múzeum, Esztergom, Hungary.

39. Gráfik, *Hajózás és gabonakereskedelem*, 46.

40. Arenstein, "Die Eisverhältnisse der Donau," 361. While his short talk included observations collected over a period of only three weeks that winter, two years later in 1850, he published a book that covered several years' worth of observations.

41. Shedel, "Mother of It All."

42. The *Vormäz* period was the era between the Congress of Vienna in 1815 and the revolution of March 1848.

43. Boyer, *Political Radicalism*, 1–5.

44. Sked, *Decline and Fall*, 82; and Okey, *Habsburg Monarchy*, 128–32.

45. "Die Zollreform in Österreich: I. Einleitung," *Austria Tagblatt für Handel und Gewerbe, öffentliche Bauten und Verkehrsmittel* (Vienna), January 2, 1851.

46. The Hungarian Ministry of Public Works and Transport promoted three areas of national development—land and water communication and traffic, engineering works, and bridge and street construction—all three of which touched on constitutive elements in the Danube's regulation.

47. Ember, "A magyarországi építészeti," 368; and Dóka, "A vízügyi szakigazgatás fejlődése, I," 525.

48. Dóka, *A vízimunkálatok irányítása*, 123–25.

49. Blackbourn, *Conquest of Nature*, 97.

50. On the Upper Danube, previous expenditures between 1818 and 1849 had amounted to 7 million florins in the thirty years, while officials committed 6.5 million florins in just the next decade (1850–61). Pasetti, *Notizen über die Donauregulierung*, 61.

51. Czoernig, *Österreich's Neugestaltung*, 324.

52. Hohensinner, "Die Verwandlung der Donau," 35.

53. Hohensinner and Schmid, "The More Dikes, the Higher the Floods," 223.

54. K.k. Direction der administrativen Statistik, *Tafeln zur Statistik . . . Jahre 1849–51*; K.k. Direction der administrativen Statistik, *Tafeln zur Statistik . . . Jahre 1852–54*; K.k. Direction der administrativen Statistik, *Tafeln zur Statistik . . . Jahre 1855–57*; and K.k. Direction der administrativen Statistik, *Tafeln zur Statistik . . . Jahre 1858–59*.

55. K.k. Direction der administrativen Statistik, *Tafeln zur Statistik . . . Jahre 1849–51*; K.k. Direction der administrativen Statistik, *Tafeln zur Statistik . . . Jahre 1852–54*; K.k. Direction der administrativen Statistik, *Tafeln zur Statistik . . . Jahre 1855–57*; and K.k. Direction der administrativen Statistik, *Tafeln zur Statistik . . . Jahre 1858–59*.

56. Anonymous, "Nem hivatolos rész," *Budapesti Hírlap*, February 8, 1855.

57. Gonda, "A magyar duna," 27–28.

58. Dóka, "A vízügyi szakigazgatás fejlődése, I," 524.

59. Dóka, *A vízimunkálatok irányítása*, 131.

60. Pasetti, *Notizen über die Donauregulierung*, 26–27.

61. Anonymous, "Amtlicher Theil," *Wiener Zeitung* (Vienna), June 21, 1850.

62. Czoernig, *Österreich's Neugestaltung*, 320–22.

63. In 1860, the total annual budget for all hydraulic engineering departments without Hungary, Transylvania, Croatia-Slavonia, or the Banat and Military Frontier was 578,272 florins, which was 528,557 in 1861, 774,727 in 1862, 657,299 in 1863, 611,466 in 1864, and 713,530 in 1865. K.k. Statistische Central-Commission, *Tafel zur Statistik . . . Jahre 1860 bis 1865*.

64. Coen, *Climate in Motion*, 281.

65. Pinke, "Modernization and Decline," 93–97.

66. Fritsch, "Eine Reise.," 34.

67. Jäckel, "Der Vögelzug und anderweitige Wahrnehmungen," 498.

68. Bastian and Bernhardt, "Anthropogenic Landscape Changes," 140.

69. R. Habsburg et al., "Zwölf Frühlingstage," 4.

70. "Vidéki tudósítók," *Pesti Napló* (Budapest), September 22, 1866.

71. Gatejel, *Engineering the Lower Danube*.

72. Danube Steam Navigation Company to Ministry of Public Works and Transport, November 4, 1867, box 17, folder 8 hajózás, K173, National Archives of Hungary, Budapest.

73. László Szilka to Ministry of Public Works and Transport, December 16, 1867, box 17, folder 8 hajózás, K173, MNL, Budapest, Hungary.

74. Torontál and Csanád County Officials to Ministry of Public Works and Transport, January 28, 1868, box 17, folder 8 hajózás, K173, National Archives of Hungary, Budapest.

75. Winckler, *Übersicht des Schiffs*, 2.

76. Dénes, *A magyar vízszabályozás*, 189.

77. In 1884, it became the Hungarian Chamber of Commerce, though there were several other Chambers of Commerce throughout Transleithania. The Pest-based chamber had between three hundred and four hundred members.

78. Suess, *Die Aufgabe der Donau*, 1.

79. *Stenographische Protokolle* (vol. 4, 1887), 3857–58.

80. Domaszewski, *Betrachtungen über die Ausnützung*; Domaszewski, *Nutzwasser- und Wasserkraftfrage*; Domaszewski, *Regulirung des Wienflusses*; Domaszewski, *Das Wasser*; and Domaszewski, *Was kostet der unvermeidliche?*

81. Hieronymi, *A közutak föntartásáról*; Hieronymi, *A közlekedés*; Hieronymi *A közmunkaügyek állami kezelése*; Hieronymi, *A budapesti Dunaszakasz szabályozása*; and Hieronymi, *Die Theissregulirung*.

82. Zels, *Die Regulirungskosten*; Zels, *Schiffahrtscanal*; Zels, *Die Selbstkosten*; Zels, *Über Wasserstrassen*; Zels, *Die Activen*; and Zels, *Versuch*.

83. The imperial and provincial authorities approved laws in 1869 and 1870, respectively, to fund the Danube's regulation at Vienna, but in 1877 and 1878, they approved another law to reauthorize funds to complete the task.

84. Another law at the same time, Law XXIII, shifted hydraulic and hydrological issues from the Ministry of Public Works and Transport to the Agriculture, Industry, and Trade Ministry (Földmivelés-Ipar- és Kereskedelmi Miniszterium). When the Agriculture Ministry became its own ministry in 1889 (Law XVIII), it retained the portfolio for all hydraulic engineering

projects, such as river regulation, irrigation and navigation canal excavation, and embankment construction.

85. These figures nevertheless paled in comparison to the Rhine's ongoing regulation costs, which amounted to 355 million marks (over 200 million florins) from 1816 to 1891. Oelwein, *Die Binnen-Wasserstrassen*, 8.

86. Oberösterreichischer Landesausschuss, *Bericht über die Thätigkeit* (1890), 177; and Oberösterreichischer Landesausschuss, *Bericht über die Thätigkeit* (1896), 143–44.

87. *Az 1892*, vol. 29, 192.

88. *Az 1892*, vol. 31, 412.

89. Paulmann, "Straits of Europe," 22–25.

90. Šedivý, "Metternich and the Suez Canal," 374.

91. Bolzano, *Leben Franz Joseph Ritters von Gerstner*, 21.

92. Loessi, *Technischer Bericht*; *Denkschrift über das Projekt eines Donau-Save-Schiffahrt-Canales*.

93. "Duna-Odera-csatorna," *Vasúti és Közlekedési Közlöny* (Budapest), June 20, 1872.

94. *Stenographische Protokolle* (vol. 2, 1873), 1659–74.

95. "A 'Budapesti Közlöny' magántáviratai," *Budapesti Közlöny* (Budapest), November 18, 1874.

96. A. Mayer, *Der Donau-Oder-Kanal als Aktien-Unternehrnung* (Waldheim, 1873); *Denkschrift über der Ausbau der Wasserstrassen in Oesterreich und über den Bau eines Donau-Oder-Canales* (Steyermühl, 1884); Viktor Rusz, *Eine Schiffahrtsstrasse Donau-Moldau-Elbe* (Konegen, 1884); A. Skene, *Der Donau-Oder-Kanal* (Frick, 1886); *Discussion über den Ausbau der Wasserstrassen in Oesterreich und insbesondere über die Herstellung eines Donau-Oder-Canals* (Frick, 1891); Geo Gothein, *Das Donau-Oder-Kanalprojekt* (Hayn, 1897); Arthur Oelwein, *Das Donau-Oder-Kanalprojekt* (Hayn, 1897); Arthur Oelwein and J. Böhm, *Das Donau-Oder-Kanalprojekt* (Hayn, 1897); Kaftan et al, *Das Donau-Moldau-Elbe-Kanalprojekt* (Hayn, 1897); J. Kaftan, F. Steiner, R. Urbanitzky, *Gegenwartiger Stand des Donau-Moldau-Elbe-Kanalprojekts* (Hayn, 1897); E. von Weber, *Das Donau-Oder-Kanalprojekt* (Hayn, 1897); S. Herczfeld, *A Duna-Odera csatorna közgazdasági jelentőségéről* (A magy. hajózási egyesület, 1898); and Kereskedelmi és iparkamara, *Bizottsági jelentés a Duna-Odera-csatorna ügyében* (Pesti könyvnyomda, 1899).

97. According to Oelwein, *Die Binnen-Wasserstrassen*, 10.

98. Projects included the Rhine-Weser, Weser-Elbe, Spree-Elbe-Oder, a lateral Oder canal, and Baltic-North Sea canals.

99. Csikvári, *A közlekedési eszközök*, 217, 235.

100. Oelwein, *Ausbau der Wasserstrassen*, 12. The currency conversion is taken from Edvinsson, "Historical Currency Converter."

101. Ruß, *Der volkswirtschaftliche Wert*, 6.

102. Grübler, *Rise and Fall of Infrastructures*, 81–3.

103. Oelwein, *Die Binnen-Wasserstrassen*, 12–13.

104. Oelwein, *Die Binnen-Wasserstrassen*, 15; and Oelwein, *Die Wasserstraßenfrage*, 10–12.

105. Oelwein, *Ausbau der Wasserstrassen*, 33.

106. Anonymous, "Közgazdaság," *Budapesti Hírlap*, June 25, 1896.

107. The Badeni language crisis erupted in 1897 when Prime Minister Kasimir Badeni attempted to placate Czech nationalists by requiring government officials in the Bohemian lands to conduct business in both German and Czech. German speakers in Bohemia castigated this decision, and the ensuing protests and political obstruction brought down Badeni's government.

108. Lindström, *Empire and Identity*, 43–44.

109. Okey, *Habsburg Monarchy*, 348.

110. A crown in the 1890s would be equivalent to approximately $24 today.

111. "Az osztrák trónbeszéd," *Pesti Napló* (Budapest), February 5, 1901; "A reichsrat ülése," *Pesti Napló* (Budapest), February 23, 1901; "Közgazdaság," *Pesti Napló* (Budapest), March 15, 1901; and "A budapesti értéktőzsde," *Pesti Napló* (Budapest), March 22, 1901.

112. "A magyar vizutak kérdése," *Pesti Hírlap* (Budapest), March 28, 1901.

113. *Stenographische Protokolle* (vol. 3, 1901), 2896.

114. "Niederösterreichischer Landtag," *Deutsches Volksblatt* (Vienna), June 21, 1901; "Niederösterreichischer Landtag," *Das Vaterland* (Vienna), June 21, 1901; "Vasutak és csatornák," *Budapesti Hírlap* (Budapest), June 12, 1901; "Regulirung der Wasserläufe," *Wiener Zeitung* (Vienna), June 27, 1901; "Unsere Wasserstraßen I.," *Pester Lloyd* (Budapest), June 19, 1901; "Unsere Wasserstraßen II.," *Pester Llyod* (Budapest), June 20, 1901; "Unsere Wasserstraßen III.," *Pester Lloyd* (Budapest), June 21, 1901; and "Ship Canals in Austria," *Geographical Journal*, vol. 18, no. 3 (1901): 289–91.

115. "Fünfter deutsch-österreichisch-ungarischer Binnenschiffahrts-Congress," *Der Bautechniker* (Vienna), September 13, 1901.

116. Coen, *Climate in Motion*, 269.

117. Oberösterreichischer Landesausschuss, *Bericht über die Thätigkeit* (1902), 244.

118. "Donau-Moldau-Canal," *Linzer Volksblatt* (Linz), October 16, 1901.

119. *Az 1901*, vol. 1, 1–3.

120. Deputy Pál Kovács, a rail engineer, gave an impassioned speech—pointing to the Netherlands and Belgium as perfect examples—stating that regulating Hungary's numerous rivers and even building new canals were the next steps in complementing its rail network, and would reduce transportation costs and encourage trade. Presaging larger debates to come, he also postulated that a Danube-Tisza Canal would both improve navigation and help irrigate and drain the land between the land's two largest rivers, *Az 1896*, vol. 27, 37.

121. *Az 1901*, vol. 2, 243; and *Az 1901*, vol. 3, 89.

122. *Az 1901*, vol. 5, 126–35.

123. *Az 1910*, vol. 1, 1–3.

124. "Wien, die Deutschen und die Schiffahrtskanäle," *Reichspost* (Vienna), August 20, 1910.

125. "Regierungsbericht über die Wasserstrassenfrage," *Neue Freie Presse* (Vienna), September 6, 1910.

126. "Die Kanalbauten in Galizien," *Neue Freie Presse* (Vienna), August 27, 1912; "Parlamentarisches," *Neues Wiener Tagblatt* (Vienna), August 28, 1912, p. 4; and *Fremdenblatt* (Vienna), August 29, 1912.

127. *Az 1910*, vol. 17, 360.

128. "Die Aenderung des Wasserstrassengesetzes," *Neues Wiener Tagblatt* (Vienna), January 18, 1911.

129. *Die Zeit* (Vienna), January 29, 1914.

130. *Az 1910*, vol. 17, 393.

131. "A meghódított Duna," *Fővárosi Hírlap* (Budapest), January 3, 1917.

132. "Die Donaukonferenz," *Pester Lloyd* (Budapest), September 4, 1916.

133. Béla Antóny, "Osztrák vizi-utak kongreszszusa Bécsben," *Esztergom* (Esztergom, Hungary), July 5, 1917.

134. Halter, "Das zentraleuropäisclie Wasserstraßennetz."

135. "The Integration of European Waterways Working Paper," *SPIN-TN*, September 10, 2004, https://www.ccr-zkr.org/files/histoireCCNR/17_the-integration-of-european-waterways.pdf.

136. "The Study Confirmed Water Corridor Danube–Odra–Elbe Stands a Chance," *Czech Ministry of Transportation*, October 19, 2018, https://www.mdcr.cz/Media/Media-a-tiskove-zpravy/Vodni-koridor-Dunaj-%E2%80%93-Odra-%E2%80%93-Labe-ma-sanci,-st-(1).

137. Alex Blackburn, "Exploring European Inland Water Freight with Maps: Visualising Transport Volumes on Rivers and Canals," *UNECE Transport*, January 18, 2023, https://w3.unece.org/Stories/2023/01/inland_waterway_freight/.

138. "Rail and Waterborne—Best for Low-Carbon Motorised Transport," *European Environmental Agency*, March 24, 2021, https://www.eea.europa.eu/publications/rail-and-waterborne-transport.

139. "Benefits of the Danube–Oder–Elbe Water Corridor," *Vodní koridor Dunaj–Odra–Labe*, https://www.d-o-l.cz/index.php/en/benefitsdoe.

140. "The Danube-Oder-Elbe Canal: Billions for a Senseless Mammoth Project," *WWF*, March 17, 2020, https://wwf.panda.org/wwf_news/?361070/Danube-Oder-Elbe-Canal.

141. Antonia Zimmermann and Hanne Cokelaere, "Europe's Dry Rivers Put Climate Ambitions at Risk," *Politico*, August 16, 2023, https://www.politico.eu/article/europe-dry-river-climate-change-ambitions-risk/.

142. Petra Sorge, Wilfried Eckl-Dorna, and Carolynn Look, "Germany's Economy Is Carried on the Rhine's Shrinking Back," *Insurance Journal*, August 2, 2023, https://www.insurancejournal.com/news/international/2023/08/02/733224.htm; and Jenny Gross, "Low Water Levels Disrupt European River Cruises, a Favorite of U.S. Tourists," *New York Times*, August 29, 2022, https://www.nytimes.com/2022/08/29/travel/river-cruises-drought-europe.html.

Chapter 3. The Danube as a People Network

1. Palacký, "Letter to Frankfurt."
2. Forgách, *Die schiffbare Donau*, 1.
3. Meissinger, *Historische Donauschiffahrt*, 39.
4. Meissinger, *Historische Donauschiffahrt*, 16.
5. *Reise auf der*, foreword.
6. Demény and Holka, *Statistics of the Centuries*, 140.
7. Hinkel, *Wien an der Donau*, 113.
8. Pasetti, *Notizen über die Donauregulierung*, 7.
9. Kállay, "Ungarischer Donauhandel," 46.
10. Planche, *Descent of the Danube*, 115.
11. Haidvogl and Gingrich, "Wasserstrasse Donau," 91–92.
12. Schultes, *Donau-Fahrten*, 144.
13. Schultes, *Donau-Fahrten*, 12.
14. Dorner, *Panorama der Österreichischen Monarchie*, 3:86.
15. Winckler, "Der Wiener Donauhandel," 12.
16. Murray, *Handbook for Travellers*, 432.
17. Bright, *Travels from Vienna*, 221; and "Hungary," 4.
18. These small advertisements were printed in the last pages of the

newspaper, where houses, businesses, and plots of land were sold, "Lizitation Curialhaus," *Pressburger Zeitung* (Bratislava), January 16, 1829.

19. Gráfik, *Hajózás és gabonakereskedelem*, 93–94.

20. Hoffman, "Die Donau und Österreich," 31.

21. *Allgemeines Reichs-Gesetz*, 643.

22. Széchenyi, *Über die Donauschifffahrt*, 14.

23. Hohensinner, "'Wie viele Fahrzeuge liegen?'"

24. Tredgold, *Remarks on Steam Navigation*, 22–23.

25. Maclean, *Remarks on the Facility*, 3–4; and Dick, *Christian Philosopher*, 410.

26. Berend, "Hungary in the Austro-Hungarian Monarchy," 8.

27. Széchenyi, *Über die Donauschifffahrt*, 16.

28. Barany, "Age of Royal Absolutism," 198.

29. Quinn, *Steam Voyage*, 88.

30. Vári, "Functions of Ethnic Stereotypes."

31. Malm, *Fossil Capital*, 200–203.

32. Gatejel, *Engineering the Lower Danube*, 106–11.

33. *Handbook for Travelers*, 292.

34. "Ob Oesterreich fortschreitet," *Der Adler* (Vienna), January 2, 1838.

35. Meissinger, *Historische Donauschiffahrt*, 39.

36. Sommer, *Das Kaiserthum Oesterreich*, 9.

37. In 1841, the DDSG's river steamers used more than 17,000 tons of bituminous coal (*Steinkohle*) alone, K.k. Direction der administrativen Statistik, "Dampfschiffahrt und Eisenbahnen," *Tafeln zur Statistik . . . Jahr 1841*, vol. 14 (1844).

38. Hajnal, *Danube*, 128.

39. B. R. Mitchell, *European Historical Statistics*, 581–82.

40. Huszár, "Zu den Beziehungen," 140.

41. A law in 1854 modernized mining practices in the empire, and total coal production increased from 30 kilotons in 1830 to 800 kilotons by 1867.

42. K.k. Direction der administrativen Statistik, "Dampfschiffahrt und Eisenbahnen," 445–53.

43. Demény and Holka, *Statistics of the Centuries*, 164.

44. István Kovács writes to István Széchenyi, June 3, 1846, Széchenyi iratok, Magyar Környezetvédelmi és Vízügyi Múzeum [Hungarian Environmental Protection and Hydrological Museum], Esztergom, Hungary.

45. K.k. Statistische Central-Commission, *Statistisches Jahrbuch . . . Jahr 1868* (1870), 127.

46. K.k. Direction der administrativen Statistik, "Werth der Waaren-Einfuhr aus Ungern und Siebenbürgen in die anderen innerhalb der Zoll-Linie befindlichen österrichischen Provinzen in den Jahren 1831 bis 1840, nebst dem Durchschnittswerthe in demselben Zeitabschnitte in Dalmantien und den quarnerischen Inseln Statt gefunden Waaren-Einfuhr," *Tafeln zur Statistik... Jahr 1841* (1844); and K.k. Direction der administrativen Statistik, "Dampfschiffahrt," *Tafeln zur Statistik der österreichischen Monarchie für die Jahre 1847 und 1848* (k.k. Hof und Staatsdruckerei, 1853).

47. K.k. Direction der administrativen Statistik, "Werth der Waaren-Einfuhr aus Ungern und Siebenbürgen."

48. K.k. Direction der administrativen Statistik, "Handel des österreichischen Zollgebiets," *Tafeln zur Statistik... Jahre 1845 und 1846*, 64.

49. K.k. Direction der administrativen Statistik, "Handel des österreichischen Zollgebiets, 34.

50. Kohl, *Die Donau*, 306.

51. "Anzeige des Stellwagens in Scheibbs," *Wiener Zeitung*, March 3, 1847.

52. "Kundmachung," *Wiener Zeitung*, January 20, 1847.

53. k.k. Direction der administrativen Statistik, "Dampfschiffahrt und Eisenbahnen," 417–27.

54. Gráfik, *Hajózás és gabonakereskedelem*, 84–85.

55. Barany, "Age of Royal Absolutism," 201.

56. Barany, "Age of Royal Absolutism," 201–8.

57. *Sitzungs-Protocoll* (1838), 11; and *Sitzungs-Protokoll* (1840), 18.

58. *Sitzungs-Protocoll* (1838), 8.

59. K.k. Direction der administrativen Statistik, "Dampfschiffahrt und Eisenbahnen," 377–88.

60. Gonda, "A magyar duna," 33–34.

61. Mittas, "Reconstruction and Transformation," 136–37.

62. River Department Royal Engineering Bureau in Becse to KMKM, April 22, 1869, box 27, folder 8 hajózás, K173, National Archives of Hungary, Budapest.

63. Smil, *Energy Transitions*, 100–145.

64. Malm, *Fossil Capital*, 1–19.

65. Coen, *Climate in Motion*, 239–73.

66. Fallon, "Passau to Pesth," 453.

67. K.k. Direction der administrativen Statistik, "Zoll Gefäll," *Tafeln zur Statistik... Neue Folge I. Das Jahr 1851* (1856), 4.

68. Interestingly enough, in May 1850, the *Leipziger Zeitung* also praised

Bruck's dissolution of Elbe customs on the Austrian stretch and claimed that he would certainly follow suit on the Danube once it had been regulated. "Handel und Industrie" *Leipziger Zeitung* (Germany), May 22, 1850. Besides customs, local authorities charged money via tolls to cross bridges or use towpaths, to rent horses along particular stretches, and for certain privileges (Wasserzoll, Zillenrecht, Zillenaufschlag, Bodenrecht, Stationszoll, Pferdemauth, and Wasser-Roßmauth). *Allgemeines Reichs-Gesetz*, 643.

69. Komlos, *Habsburg Monarchy*, 1–13.

70. Komlos, *Habsburg Monarchy*, 1–13.

71. In 1846–50, the DDSG transported 130,000 tons of goods on average per year, whereas in 1851–55, this was 560,000 tons on average per year.

72. K.k. Statistische Central-Commission, *Tafeln zur Statistik . . . Jahre 1860 bis 1865*, 21–22; Tirscher, "A hajózás és forgalom," 226.

73. K.k. Statistische Central-Commission, *Statistisches Jahrbuch . . . Jahr 1877*, vol. 4 (1880), 32.

74. Winckler, *Übersicht des Schiffs- und Waarenverkehrs*, 7.

75. Tirscher, "A hajózás és forgalom," 218.

76. Wallace, *Auf der Donau*, 13.

77. Certain Habsburg advisers, such as Metternich, and later Bruck and Schwarzenberg, had advocated for tariff reductions on German goods or even on entry to the *Zollverein*, in order to counter Prussia's influence in the German Confederation and stimulate economic growth. Good, *Economic Rise*, 80–82.

78. "Ermächtigung," *Verordnungs-Blatt*, 445.

79. Okey, *Habsburg Monarchy*, 169.

80. Tirscher, "A hajózás és forgalom," 220.

81. K.k. Statistische Central-Commission, *Statistisches Jahrbuch*, 145; and Gráfik, *Hajózás és gabonakereskedelem*, 93, 98.

82. Tirscher, "A hajózás és forgalom," 219; and Arthur Schott, "Die drei Banater Städte Pantschowa, Werschitz und Orschowa," *Das Ausland*, June 15, 1849, p. 569.

83. Okey, *Habsburg Monarchy*, 169.

84. "Donau-Dampfschiffahrt,"*Austria Tagblatt für Handel und Gewerbe, öffentliche Bauten und Verkehrsmittel* (Vienna), February 17, 1851.

85. *Geschäfts-Bericht der Betriebs-Direction*, 52.

86. DDSG to KMKM, November 4, 1867, box 17, folder 8 hajózás, K173, National Archives of Hungary, Budapest.

87. *Magyar Statistikai Évkönyv*, 185.

88. Gráfik, *Hajózás és gabonakereskedelem*, 99.

89. K.k. Statistische Central-Commission, "Waarenverkehr zwischen Oesterreich," 159.

90. As quoted in Gráfik, *Hajózás és gabonakereskedelem*, 99.

91. Rail and River Inspectorate's office writing to Trade Ministry, September 2, 1894, folder 6, K228, National Archives of Hungary, Budapest.

92. "Schiffverkehr auf der oberen Donau," 34.

93. K.k. Statistische Central-Commission, *Statistik des Verkehrs*, 1–15.

94. Meissinger, *Historische Donauschiffahrt*, 38.

95. K.k. Statistische Central-Commission, "Dampfschiffahrt," *Tafeln zur Statistik... Jahre 1860 bis 1865*, vol. 2 (1871), 12.

96. "Köztudomású, hogy ezen vállalat forgalmának tetemes része közvetlenül az Ausztria és Magyarország közti forgalmare vonatkozik, miután a Duna a monarchia két felének természetes összefüzője," *Az 1875-ik évi augustus hó 28-ára kihirdetett országgyűlés főrendi házának irományai*, vol. 8 (Pesti Könyvnyomda-Részvény-Társulat, 1878), 165.

97. K.k. Statistische Central-Commission, *Statistik des Verkehrs... Jahre 1894 und 1895*, xliii–xliv.

98. Brushes and sieves went up over 1,000 percent; flax, hemp, rubber, and rosin nearly 300 percent; and foodstuffs nearly 250 percent. *Statistik des Verkehrs in den im Reichsrathe vertretenen Königreichen und Ländern vornehmlich für die Jahre 1894 und 1895*, xli–xlii.

99. K.k. Statistische Zentral-Commission, "Der Zwischenverkehr der im Reichsrate... Jahren 1900 und 1901" (1905), lxx; K.k. Statistische Zentral-Commission, "Der Zwischenverkehr der im Reichsrate... Jahren 1902 bis 1905" (1905), lxvii; K.k. Statistische Zentralkommission, *Österreichisches Statistisches Handbuch* 1908), 327; K.k. Statistische Zentralkommission, "Der Zwischenverkehr der im Reichsrate... 1906 und 1907" (1911), lv–lvii.

100. Komlos, *Habsburg Monarchy*, 10.

101. This was not a clear-cut "rail vs. water" rivalry, as both the Hungarian and Austrian State Railway Companies ran passenger lines on the Danube as well. Erste k.k. priv. Donau-Dampfschiffahrts-Gesellschaft, *Geschäfts-bericht* (1889).

102. *Az 1910*, vol. 7, 395–96.

103. Jungwirth et al., *Österreichs Donau*, 185–212; Landgraf, "Halászat"; Anonymous, "A vadászat," *Budapesti Hírlap*, October 26, 1891; István Chernel, "Az alvidéki szárnyas ínségesek érdekében," *Vadászlap*, August 5. 1892.

104. *Österr.-ungar. Blätter für Geflügel*, 205–12.

105. Gruffat, "Beautiful Public Danube."

106. Anonymous, "Die Verwendung von Fischen zur Verpflegung des Soldaten," *Der Militärarzt*, September 9, 1904.

107. Gottlieb, *Die internationale Fischerei-Conferenz*, 7–20.

108. *Österr.-ungar. Blätter für Geflügel*, 205–12.

109. Anonymous, "Halászati érdekek," *Budapesti Hírlap*, April 3, 1894.

110. K.k. Statistische Zentralkommission, "Gesamtverkehr," x–xi.

111. These data are aggregated from the statistical reports (*évkönyvek*) from the Central Statistical Office in Budapest.

112. K.k. Statistische Central-Commission, *Österreichisches Statistisches Handbuch . . . 1914*, 165.

113. *Österreichisches Statistisches Handbuch 1914*, 167.

114. This was not homogeneous across the empire. In the period 1869–1910, the Austrian core lands' urban population proportion leaped from 21.2 percent to 37 percent, in the Bohemian lands from 6.7 percent to 18.7 percent, in Galicia/Bukovina from 5.8 percent to 12.7 percent, and in Transleithania from 10.5 percent to 17.4 percent. Fassmann, "City-Size Distribution," 11.

115. The four Danube cities in the top twenty were Vienna (1), Budapest (2), Linz (13), and Pressburg/Poszony (16). The nine cities on Danube tributaries were Graz (5) on the Mur, Brünn (8) on the Svratka/Svitava confluence, which flows into the Thaya and later Moravia River, Szeged (9) on the Tisza, Szabadka (10) on the Tisza, Czernowitz (12) on the Prut, Zagreb (15) on the Save, Temesvár (17) on the Bega Canal, Nagyvárad/Großwardein (19) on the Sebes-Körös, and Arad (20) on the Maros. Admittedly, not all these cities used their local rivers to the same extent as commercial paths to the Danube, but most certainly for the sorts of daily practices that existed on the Danube.

116. Gierlinger et al, "Feeding and Cleaning the City," 224–25.

117. Donau-Regulirungs-Commission, *Beschreibung der Arbeiten*, 6.

118. Wallace, *Auf der Donau*, 26.

119. Winckler, "Übersicht des Schiffs- und Waarenverkehrs," 10.

120. K.k. Statistische Central-Commission, *Österreichisches Statistisches Handbuch . . . 1897*, 225; K.k. Statistische Central-Commission, ed., *Österreichisches Statistisches Handbuch . . . 1907*, 337; Gierlinger et al., "Feeding and Cleaning the City," 225.

121. Mokre, "Environs Map," 91, 93.

122. Gierlinger et al., "Feeding and Cleaning the City," 228.

123. Gráfik, *Hajózás és gabonakereskedelem*, 93–94.

124. Franz Badics, "Die Bácska," 592.

125. K.k. Statistische Central-Commission, *Oesterreichisches Statistisches Handbuch* ... 1887, 179.

126. Winckler, "Übersicht des Schiffs- und Waarenverkehrs," 35, 37, 39.

127. K.k. Statistische Central-Commission, ed., *Statistisches Jahrbuch* ... 1869, 144.

128. Szeged Steam Navigation Company to Public Works and Transportation Minister, March 11, 1867, box 27, folder 8 hajózás, K173, National Archives of Hungary, Budapest.

129. In comparison with goods leaving Szeged by train and cart, the following goods departed by ships: 96 percent of the exported wheat (24,600 tons), 99 percent of barley (9,940 tons), 99 percent of corn (1,270 tons), 45 percent of potatoes (4,580 tons), 61 percent of wood (28,300 tons), 92 percent of glass (5,050 tons), and 44 percent of tobacco (1,430 tons). Gráfik, *Hajózás és gabonakereskedelem*, 92.

130. DDSG to Interior Minister Taaffe, December 3, 1890, Inneres MdI Allgemein A 459, Vienna, Austria.

131. The company decided to halt its Black Sea passenger traffic and reduce certain stretches on the Lower Danube as well. Erste k.k. priv. Donau-Dampfschiffahrts-Gesellschaft, *Geschäfts-Bericht* (1897), 7.

132. Oberösterreichischer Landesausschuss, Bericht über die Thätigkeit (1896), 141–43.

133. *Geschäfts-Bericht für das Jahr 1896*, 2.

134. Oberösterreichischer Landesausschuss, *Bericht über die Thätigkeit* (1896), 141–43.

135. *Rechenschaftsbericht des Gemeinderathes der Landeshauptstadt Linz über seine Thätigkeit im Jahre 1901 nebst anderen statistischen Daten* (Selbstverlag des Gemeinderathes, 1902), 138, 484.

136. Melk municipal council meeting, October 9, 1895/February 15, 1896, *Gemeinderathsprotokolle*, Melk, Austria.

137. Melk municipal council meeting, April 30, 1892, *Gemeinderathsprotokolle*, Melk, Austria.

138. Franz Xaver Linde, *Chronik des Marktes*, 378–416.

139. Erste k.k. priv. Donau-Dampfschiffahrts-Gesellschaft, *Geschäfts-Bericht* (1868), 11.

140. Vari, "From 'Paris of the East.'"

141. "Az őstörténelmi nemzetközi kongresszusra," *Vasárnapi Újság*, May 28, 1876.

142. *Geschäfts-Bericht für das Jahr 1896*, 1–2.

143. Oberösterreichischer Landesausschuss, *Bericht über die Thätigkeit* (1902), 243.

144. Along with Aggsbach, the communities included Groisbach, Willendorf, Köfering, Thalheim, Giesshübl, Wiemannsreith, Leeb, Hublhof, Hof, Letzendorf, Schlanbing, Maria Laach, Zeising, Friedensdorf, Loitzendorf, Nonnensdorf, Zintring, and Grimsing. Mayor of Aggsbach to DDSG, June 3, 1902, AT-OeStA/AVA Handel HMallg A 914 Donauangelegenheiten (K/b), Zl. 25001-Ende.

145. Landesverband für Fremdenverkehr in Oberösterreich to the k.k. Trade Ministry, May 10, 1902, AT-OeStA/AVA Handel HMallg A 914 Donauangelegenheiten (K/b), Zl. 25001-Ende.

146. Alkoven's mayor to Upper Austria *Statthalter*, November 24, 1902, AT-OeStA/AVA Handel HMallg A 914 Donauangelegenheiten.

147. Oberösterreichischer Landesausschuss, *Bericht über die Thätigkeit . . . Sommer 1902 bis Herbst 1908*, 188.

148. Aggregated data from Annual Statistical Reports (*Statisztikai évkönyvek*).

149. *Schriften der in Budapest am 4. September des Jahres 1916 abgehaltenen Donaukonferenz* ("Pátria" literar. Unternehmen und Druckerei Aktien-Gesellschaft, 1916), 69.

150. *Rechenschaftsbericht des Gemeinderathes der Landeshauptstadt Linz* (1908), 157.

151. Aggregated data from Annual Statistical Reports (*Statisztikai évkönyvek*).

152. Heksch, *Illustrirter Führer*, v–vi.

153. McNeill and McNeill, *Human Web*, 219–21.

154. Gingrich et al., "Danube and Vienna," 292–94.

155. Berend and Ránki, "Economic Problems."

Chapter 4. Overcoming Danubian Dangers

1. Otto Grad von Fünfkirchen, "Der Bretterwandbach in Windisch-Matrei," *Innsbrucker Zeitung*, March 6, 1850, p. 229.

2. Anonymous, *Die Schreckens-Tage*.

3. Anonymous, "Privatgedanken eines Kameralisten über Finanzwesen: Schluß der IV. Abhandlung Oekonomie B. Weise Sparsamkeit," *Innsbrucker Zeitung*, August 24, 1849, p. 799.

4. Fagan, *Little Ice Age*.

5. Parker, *Global Crisis*, xxi–xxix.

6. Brönnimann et al., "Climate from 1800 to 1970," 312–13.

7. Strickland and Church, "Leibnitz's Observations," 526–29.

8. Lóczy, "Danube," 237.

9. "Hydrotechnisches," *Wiener Theater-Zeitung*, April 22, 1830.

10. Pfister et al., "Early Modern Europe," 276–81.

11. Hohensinner, "Historische Hochwässer,"10.

12. Rácz, *Steppe to Europe*, 225.

13. Lászlóffy-Böhm, "A Tiszavölgy," 111.

14. Giosan et al., "Early Anthropogenic Transformation," 5; and Glassl, "Der Ausbau der ungarischen Wasserstraßen," 54.

15. Hohensinner and Schmid, "The More Dikes, the Higher the Floods," 222–25.

16. "Kundmachung," *Wiener Zeitung*, January 29, 1829.

17. Hohensinner and Hahmann, "Historische Wasserbauten,"14–15.

18. Fejér, *Árvizek és belvizek szorításában*, 17.

19. Wall text, Main Hall, Danube Museum, Esztergom, Hungary.

20. Sárközi, *Árvizek, ármentesítés és folyószabályozás*, 18–19.

21. Gunst, *Agrarian Development*, 13.

22. Hohensinner, Lager et al., "Changes in Water and Land," 160.

23. Gonda, "A magyar duna," 25.

24. Sárközi, *Árvizek, ármentesítés és folyószabályozás*, 17–18.

25. Ball, *Water Kingdom*, 21, 167–78.

26. Cioc, *Rhine*.

27. Doyle, *Source*, 28–34.

28. "Pressburg," *Pressburger Zeitung*, April 20, 1830.

29. Károlyi, "A magyar vízi munkálatok," 64.

30. Sándor Somogyi, *A XIX. századi folyószabályozások és ármentesítések földrajzi és ökológiai hatásai Magyarországon* (Budapest, 2000), 157, 247, quoted in Márton Simonkay, "Felső-Duna-völgy dualizmuskori szabályozása," (bachelor's thesis, Eötvös Loránd University, 2013), 4.

31. "1840. évi IV. Törvénycikk a Duna s egyéb folyamok szabályozásáról," https://net.jogtar.hu/ezer-ev-torveny?docid=84000004.TV.

32. Károlyi, "A magyar vízi munkálatok," 104.

33. Benedict Anderson's formulation focused on vernacular "print capitalism" and implied that only literate communities took part in the rituals of reading papers and creating a joint sense of time and location, but the spread of news in the Habsburg Empire also followed informal networks such as book clubs and coffee houses, as well as the role that pastors and others served in disseminating news to illiterate groups.

34. Anonymous, "Aufruf an edeldenkende, wohlthätige Menschen," *Wiener Zeitung*, March 4, 1830.

35. Rohr, "Danube Floods"; Mauelshagen, "Disaster and Political Culture," 49–55; Glaser et al., "Variability of European Floods"; and Berry, "Great Tyne Flood."

36. Sartori, *Wien's Tage der Gefahr*, 107–32.

37. Anonymous, "Einladung an die edlen und menschenfreundlichen Bewohner Wiens," *Wiener Zeitung*, March 5, 1830.

38. Anonymous, "Wien, den 2. März," *Brünner Zeitung der k.k. priv. mähr. Lehenbank*, March 5, 1830; Anonymous, "Pressburg, 4tn März," *Presssburger Zeitung*, March 5, 1830; Anonymous "Wien, den 2. März," *Brünner Zeitung der k.k. priv. mähr. Lehenbank*, March 6, 1830; and Anonymous, "Wien," *Grazer Zeitung*, March 6, 1830.

39. Heliser, *Rövid tudósítás*, 7–19.

40. Judson, *Habsburg Empire*, 137–45.

41. Brázdil et al., "Historical Floods," 124–25.

42. Anonymous, "Für das Stadtgespräch," *Die Gegenwart. Politisch-literarisches Tagblatt*, November 11, 1845, p. 168.

43. Anonymous, "Das Inland," *Die Gegenwart. Politisch-literarisches Tagblatt*, January 4, 1847, p. 9.

44. Neundlinger, "Disaster Ahead."

45. James B. Calvert, "The Electromagnetic Telegraph," http://mysite.du.edu/~jcalvert/tel/morse/morse.htm#G.

46. "Über Zusammenstellung der gleichseitigen Wasserstandsbeobachtungen an sämmtlichen schiffbaren Gewässern der österreichischen Monarchie,"*Austria Tagblatt für Handel und Gewerbe, öffentliche Bauten und Verkehrsmittel*, February 10, 1851.

47. Coen, *Climate in Motion*.

48. Hammerl, "Viennese School of Climatology."

49. "Geschichte."

50. Hegyfoky, "Wasserstand der Flüsse."

51. Anonymous, "Az árviz pusztitása," *Budapesti Hírlap*, May 29, 1907, p. 10.

52. "Tages-Neuigkeiten," *Deutsches Volksblatt* (Vienna), February 6, 1893.

53. Oberösterreichischer Landesausschuss, *Bericht über die Thätigkeit*, 1902, 243; Oberösterreichischer Landesausschuss, *Bericht über die Thätigkeit*, 1908, 197; and "Was ist die Hydrografie Österreichs?"

54. Osterhammel, *Transformation of the World*, 614.

55. Okey, *Habsburg Monarchy*, 128–32.
56. Sked, *Decline and Fall*, 76–77.
57. Sked, *Decline and Fall*, 82.
58. Schediwy, *Städtebilder*, 321.
59. F. Vadas, "Die Regulierung der Donau," 81.
60. József Péch, "Dunaszabályozási Levelek I," *Pesti Napló*, December 24, 1858; József Péch, "Dunaszabályozási Levelek III," *Pesti Napló*, December 25, 1858; József Péch, "Dunaszabályozási Levelek IV," *Pesti Napló*, December 28, 1858; József Péch, "Dunaszabályozási Levelek V," *Pesti Napló*, December 29, 1858; and József Péch, "Dunaszabályozási Levelek VI," *Pesti Napló*, December 31, 1858.
61. Pasetti, *Notizen über die Donauregulierung*, 21.
62. Czoernig, *Oesterreich's Neugestaltung*, 323–24, 334.
63. Pasetti, *Notizen über die Donauregulierung*, 21.
64. Hofman, *Die Ueberschwemmung*, 3.
65. Anonymous, "A főrendiház üléése február 28-án.," *Pesti Napló*, March 1, 1871.
66. Dóka, "A vízügyi szakigazgatás fejlődése, II," 55.
67. Nagy, *Az 1876 évi árvízek*, 22–26.
68. "Die Eventualität einer Ueberschwemmung," *Die Presse*, January 13, 1876.
69. "Vom Eisstoß," *Neues Wiener Tagblatt*, January 19, 1876.
70. Wex, *Lecture*, 8.
71. Hieronymi, *A budapesti Duna-szakasz szabályozása*, 3.
72. Hohensinner and Schmid, "The More Dikes, the Higher the Floods," 222–24.
73. Brunner and Schneider, *Umwelt Stadt*, 312; and *Rechenschaftsbericht des Gemeinderathes der Landeshauptstadt Linz über seine Thätigkeit im Jahre 1899 nebst anderen statistischen Daten* (Selbstverlag des Gemeinderathes, 1900), 167.
74. Oberösterreichischer Landesausschuss, *Bericht über die Thätigkeit . . . 1884 . . . 1890* (1890), 179; and Oberösterreichischer Landesausschuss, *Bericht über die Thätigkeit . . . 1890 . . . 1896* (1896), 144.
75. *Statistischer Bericht*, 16.
76. "Die älteste Versicherung Tirols und ihre Vorgeschichte," *Tirol.gv.at*, https://www.tirol.gv.at/fileadmin/themen/kunst-kultur/landesarchiv/downloads/versicherung.PDF.
77. "Bemerkungen zu dem Gesetzentwurf, betreffend die Beitragsleis-

tung des Staatsschatzes zu den Kosten der Regulirung des Gailflusses in Kärnten," *Stenographische Protokolle über die Sitzungen des Hauses der Abgeordneten des österreichischen Reichsrathes, VIII. Session*, http://alex.onb.ac.at/cgi-content/alex?aid=spa&datum=0008&page=18889&size=45.

78. Gráfik, *Hajózás és gabonakereskedelem*, 10.

79. "Nichtamlichter Theil," *Wiener Zeitung*, November 7, 1899.

80. "Nichtamtlicher Theil," *Wiener Zeitung*, May 24, 1900.

81. "Kleine Chronik," *Wiener Zeitung*, February 14, 1906; and "Kleine Chronik," *Wiener Zeitung*, February 18, 1906.

82. "Kleine Chronik," *Wiener Zeitung*, September 29, 1906.

83. Oberösterreichischer Landesausschuss, *Bericht über die Thätigkeiten* (1896), 250.

84. "Nichtamtlicher Theil," *Wiener Zeitung*, April 28, 1900.

85. *Az 1906*, vol. 21, 55.

86. An additional forty companies existed in the Tisza River Valley, which reclaimed over 2.5 million hectares of land, built over 3,500 kilometers of embankments, and constructed over 9,000 kilometers of most drainage canals. Dénes, *A Magyar vízszabályozás*, 388–89.

87. Borsos, "Rivers, Marshes, and Farmlands," 205.

88. Andrásfalvy, *A Sárköz és a környező Duna*.

89. Romsics, "Peasantry," 114.

90. Kreuter, *Praktisches Handbuch*, 30–32.

91. "The Revolt of Peasants in Eastern Slovakia in 1831," *ARDSystem*, http://www.qrlink.sk/new/en/rolnicke-povstanie-na-vychodnom-slovensku-1831/.

92. "The Memorial of the Revolt of Peasants in Eastern Slovakia," *Monuments of Remembrance*, http://monuments-remembrance.eu/en/panstwa/slowacja-2/291-the-memorial-of-the-revolt-of-peasants-in-eastern-slovakia.

93. Solymos, *Élet a Dunán*, 7.

94. Andrásfalvy, "Die traditionelle Bewirtschaftung."

95. Prónay, *Skizzen aus dem Volksleben*, 51–54.

96. Anonymous, "Budapesti halászati kongresszus II," *Halászat*, April 1, 1901.

97. Ruprecht, "Bericht über die Entwässerungsarbeiten," 40–41.

98. A magyarországi cs. k. helytartóság, "Utasítás a hátóságok számára, a jégzajlás általi vizárveszélynél Budapesten," August 29, 1860, D203, box 2, folder 2, National Archives of Hungary, Budapest.

99. "Erlass des Statthalters für Niederösterreich vom 22. December 1851, Nr. 42.942," *Landesgesetz- und Regierungsblatt*, 25–26.

100. Anonymous, "Wien ist gerüstet," *Wiener Presse*, January 22, 1868, p. 9

101. Anonymous, "A Duna áradása," *Vásárnapi Újság*, January 30, 1876.

102. Anonymous, "Vorkehrungen gegen eine Ueberschwemmung," *Neue Freie Presse*, January 12, 1876.

103. Anonymous, "Ueberschwemmung in Wien," *Die Presse*, February 19, 1876.

104. Anonymous, "Feuerwehr als Wasserwehr," *Kremser-Feuerwehrzeitung*, February 27, 1886, p. 1.

105. *Rechenschaftsbericht des Gemeinderathes der Landeshauptstadt Linz über seine Thätigkeit im Jahre 1899*, 166.

106. *Rechenschaftsbericht des Gemeinderathes der Landeshauptstadt Linz … 1913*, 122.

107. "Feuilleton," *Wiener Zeitung* (Vienna), February 3, 1862.

108. Herold, *Brigittenau*, 34.

109. "Wien, 3. Februar," *Wiener Zeitung* (Vienna), February 3, 1862. His parents Archduke Franz Carl and Archduchess Sophie also visited Neubau to donate 400 florins from their own private coffers.

110. The Kingdom of Hungary's borders were essentially enclosed within the basin, so this signified a high proportion of flooded land in the Hungarian territories. Bódi et al., "Review of Historic Floods."

111. Neweklowsky, "Die Donau bei Linz," 179–80.

112. *Protokolle über die Sitzung des Hauses der Abgeordneten des Reichsrathes in der Zeit vom 29. April 1861 bis 16. December 1862* (k.k. Hof- und Staatsdruckerei, 1862), 2144; *Protokolle über die Sitzung des Hauses der Abgeordneten des Reichsrathes in der Zeit vom 29. April 1861 bis 16. December 1862* (k.k. Hof- und Staatsdruckerei, 1862), 2260.

113. Oberösterreichischer Landesausschuss, *Bericht über die Thätigkeit* (1902), 177.

114. Oberösterreichischer Landesausschuss, *Bericht über die Thätigkeit* (1902), 86–88.

115. James C. Scott castigates such high modernist projects, pointing to the tendency to oversimplify problems and solutions, which invariably fail to account for the intricacies of complex systems. Scott, *Seeing Like a State*.

116. "Die Au nach dem Hochwasser," *Nationalpark Donau Auen*, https://www.donauauen.at/wissen/natur-wissenschaft/die-donau/die-au-nach-dem-hochwasser.

117. "Lower Danube Green Corridor: Floodplain Restoration for Flood Protection," *Climate-ADAPT*, June 7, 2016, https://climate-adapt.eea.europa.eu/en/metadata/case-studies/lower-danube-green-corridor-floodplain-restoration-for-flood-protection.

118. Klein and Zellmer, *Mississippi River Tragedies*; and Cioc, *Rhine*.

119. Tóth et al., "Hortobágy Puszta."

120. Szász, *Die ungarische Landwirtschaft*, 94.

Chapter 5. Act Locally, Think Imperially

Epigraph: Jókai, "Bevezetés," 3.

1. Collins et al., "Pittsburgh's Three Rivers."

2. Jókai, "Bevezetés," 3.

3. Gyáni, "Das Verhältnis von Urbanisation"; and Gingrich et al., "Danube and Vienna."

4. Friedrich Hauer utilizes historical hydromorphological data, which the Viennese Danube projects URBWATER and ENVIEDAN have compiled, and combines that information with district demographic data to show how certain populations, such as in Leopoldstadt, grew and industrialized as the Danube's alluvial floodplain at Vienna was transformed into a few narrow channels for transportation. Hauer, "Wien und die Donau(auen)," 125–29. Statistics about Donaustadt demonstrate that the district population increased from 3,608 people in 1869 to 26,833 by 1910, a more than eightfold increase, which still amounted to only 1.3 percent of Vienna's total population. "DONAUSTADT 22. Wiener Gemeindebezirk," *wien.at*, https://www.wien.gv.at/statistik/pdf/bezirke-im-fokus-22.pdf.

5. Steidl et al., "Relations," 73.

6. Steidl et al., "Relations," 73.

7. Jekelfalussy, *Millennium of Hungary*, 378.

8. Csorba, "Transition," 70–72.

9. Watznetter, "Danube Islanders," 92.

10. Both Gary Cohen and Robert Nemes challenge the historiographical assumption that national divisions were monolithic by exploring the spaces and activities that enabled polylingual and national actors to partake in cross-cultural interactions. Cohen, "Cultural Crossings"; and Nemes, *Once and Future Budapest*.

11. Wood, "Not Just the National," 260.

12. Schweiger-Lerchenfeld, *Die Donau*, 483.

13. Vadas and Ferenczi, "Small Urban Waters"; and Matis and Bachinger, "Österreichs industrielle Entwicklung," 225.

14. Dezsényi and Hernády, *A magyar hajózás története*, 58.

15. K.k. Direction der administrativen Statistik, "Dampfschiffahrt und Eisenbahnen," *Tafeln zur Statistik... Jahr 1841* (1844), 535–46.

16. Buffe, *Marines du Danube*, 216.

17. Wunderl, "Die Geschichte der Schiffswerft Korneuburg," 28.

18. Hauer, "Wien und die Donau(auen)," 125.

19. Gierlinger et al., "Feeding and Cleaning the City," 224–25.

20. July 26, 1870, Proceedings of the Metropolitan Board of Public Works, 1870–1948, box 67, II.1.a, Folio 1, Budapest City Archives, Budapest, Hungary.

21. Reitter, *Donau-Regulirung zwischen Pest und Ofen*, 4.

22. "Hajózható csatorna terve Budapest körül," *Huszadik Század*, http://www.huszadikszazad.hu/1912-november/gazdasag/hajozhato-csatorna-terve-budapest-korul.

23. Pilsitz, "Die Donau," 195.

24. Rozsnyai, "Industrial Buildings and Halls," 639.

25. Czoernig, *Oesterreich's Neugestaltung*, 321; and Reitter, *Donau-Regulirung zwischen Pest und Ofen*, 69.

26. Schmid et al., "Danube."

27. Several excellent works explore this in the American context, including Joel A. Tarr's *The Search for the Ultimate Sink*. Bill Luckin, Dieter Schott, and Geneviève Massard-Guilbaud's edited volume, *Resources of the City*, has also compared the environmental impact of urban practices in European cities. Dieter Schott has a useful overview of urban environmental historiography, "Urban Environmental History."

28. Discussions of the "national body" differed from country to country as did the consequences presented by ideologues warning about the need for reforms. The gymnastics organization Sokol in the Bohemian lands functioned as an ersatz national organization, emphasizing a "strong mind in a strong body" for Slavs without a national state. In *The Dreyfus Affair and the Crisis of French Manhood*, Christopher E. Forth writes about how defenders of French assimilationist policies used the Dreyfus Affair to draw comparisons between physical maladies and weaknesses and the dangers of a moral weakness in the body politic. The twentieth century also experienced the extreme manifestations of eugenics discussions and the need for a "pure" national body, seen in pogroms, ethnic cleansing, and later genocidal programs. See Hull, *Absolute Destruction*; Naimark, *Fires of Hatred*; Bjørnlund, "A Fate Worse Than Dying"; and Ekmekcioglu, "A Climate for Abduction."

29. Although Helmut Gruber writes about interwar Austria, he never-

theless reveals the concern that Social Democratic leaders had for the physical and moral education of workers in Vienna. Gruber concludes, however, that such paternalistic treatment could not convince workers to change certain "immoral" or "unhealthy" practices in order to turn toward the Socialist utopia that political leaders advocated. Gruber, *Red Vienna*.

30. As quoted in Hakkarainen, *Comical Modernity*, 242.
31. Krejci, *Expedition*, 128–44.
32. Géra, "Wechselwirkung," 68.
33. Göőz, *Budapest története*, 157.
34. Jungwirth et al., *Österreichs Donau*, 141–47.
35. In Ferencváros, residents had six times higher rates of contracting cholera than residents in the center of town, and Kőbánya had twenty times higher rates. Demény and Holka, *Statistics of the Centuries*, 85.
36. Juhász, *A csatornázás története*, 40–43.
37. Hagen and Hauer, "Hygiene und Wasser."
38. Attilio Rella, "Die Assanirung der Städte in Oesterreich-Ungarn, 1848–1898," *Zeitschrift des Ingenieur- und Architekten-Vereines*, April 28, 1899, pp. 276–78.
39. Juhász, *A csatornázás története*, 58–63.
40. Juhász, *A csatornázás története*, 43.
41. Juhász, *A csatornázás története*, 70.
42. *Amts-Blatt*, 143–44.
43. Smith, "Water and Death," 172.
44. Rella, "Die Assanirung der Städte."
45. Juhász, *A csatornázás története*, 69.
46. "Circulare von der k.k. ni. ost. Landesregierung im Erzherzogthume Oesterreich unter der Enns," *Wiener Zeitung*, August 19, 1807.
47. Eder, *Bade- und Schwimmkultur*, 90, 124.
48. Eder, *Bade- und Schwimmkultur*, 135.
49. Hinkel, *Wien an der Donau*, 44.
50. Hartwell, "Public Baths in Europe," 477–78.
51. Hinkel, *Wien an der Donau*, 47.
52. Hauer, "Wien und die Donau(auen)," 124.
53. Badaeker, *Southern Germany and Austria*, 225.
54. J. Szabó, "Budapester Stadtpanorama," 138.
55. László Novotny, "Gróf Széchenyi István a magyar evezés és kajak-kenu sport megalapítója," *Széchenyi Forum*, http://www.szechenyiforum.hu/9/index.php?n=5&tartalom_id=6688&print=1.
56. Fejér, *Vizeink krónikája*, 107.

57. Anonymous, "Fővárosi hírek," *Fővárosi Lapok*, January 17, 1871.

58. Hauer and Spitzbart-Glasl, "Nebenvorteile und Erbschaften."

59. Otruba, "Linz, seine neue Strafanstalt," 303.

60. Lackner and Stadler, *Fabriken in der Stadt*, 99–100.

61. Mayrhofer and Katzinger, *Geschichte der Stadt Linz*, 81.

62. Gonda, *Die ungarische Schiffahrt*, 6–8.

63. Gráfik, *Hajózás és gabonakereskedelem*, 72.

64. K.k. Direction der administrativen Statistik, "Dampfschiffahrt und Eisenbahnen," *Tafeln zur Statistik . . . Jahr 1841* (1844), 535–46.

65. While the company had expanded its lucrative Lower Danube and Black Sea commerce by 1833, the Russian and Ottoman governments successfully challenged their presence, so by 1844, the company had eventually sold its Black Sea passenger lines to the Austrian Lloyd, a sea navigation company. Hajnal, *Danube*, 128.

66. Mayrhofer and Katzinger, *Geschichte der Stadt Linz*, 83.

67. István Széchenyi to the Győr Free Royal City Council, March 14, 1846, Széchenyi iratok 28-17.117, Magyar Környezetvédelmi és Vízügyi Múzeum [Hungarian Environmental and Hydrological Museum], Esztergom, Hungary.

68. Széchenyi had worked assiduously since the early 1830s to both regulate the Danube and spread steam navigation, and in the early 1840s was serving as the copresident of the Commission for Communication in Hungary. Joseph Voigt to István Széchenyi, March 24, 1846, Széchenyi iratok 28-17.117, Magyar Környezetvédelmi és Vízügyi Múzeum [Hungarian Environmental and Hydrological Museum], Esztergom, Hungary.

69. Writing a history of Hungarian navigation in 1899, the engineer Béla Gonda also wrote that "the city Győr was the most important trade center on the Danube . . . which even pushed the capital into the background." Gonda, *Die ungarische Schiffahrt*, 6–8.

70. Bay, *A Győri Lloyd városáért*, 19.

71. Bay, *A Győri Lloyd városáért*, 19.

72. "Napló," *Győri Közlöny*, December 16, 1858.

73. "Napló," *Győri Közlöny*, April 10, 1859.

74. K.k. Direction der administrativen Statistik, "Dampfschiffahrt und Eisenbahnen," *Tafeln zur Statistik . . . Jahr 1842* (1846), 445–53; and "Napló," *Győri Közlöny*, October 18, 1860.

75. Bay, *A Győri Lloyd városáért*, 43, 57.

76. "Napló," *Győri Közlöny*, September 15, 1861.

77. Lackner and Stadler, *Fabriken in der Stadt*, 19.

78. Károly Vörös, "Győr és Pest Harca a Dunai Gabonakereskedelemért, 1850–1881," *Arrabona* 7 (Győri Xántus János Múzeum, 1965): 471–72.
79. Jankó, *A magyar dunai gőzhajózás*, 89–90.
80. Bay, *A Győri Lloyd városáért*, 163.
81. Bana, *Győr*, 190
82. Győr Város Tanácsának Iratai 1883 I. 200, Győr Megyei Levéltár, Győr, Hungary.
83. Győr szabad s királyi város közgyülési jegyzőkönyve, Győr Megyei Levéltár, Győr, Hungary.
84. Simonkay, "'Vagy ilyen szabályozás less,'" 2–3.
85. Győr Város Tanácsának Iratai 1883.
86. *Fortsetzung der Actenstücke und Verhandlungen seit dem Jahre 1882*, 24.
87. *Rechenschaftsbericht... im Jahre 1882*, 116.
88. The DRC was established by an 1868 law that mandated the Danube's regulation near Vienna. After the Suez Canal opened in 1869, the French company sent its excavators to Vienna to start work, which took place in 1870–75.
89. *Rechenschaftsbericht... im Jahre 1882*, 120.
90. *Rechenschaftsbericht... im Jahre 1882*, 64.
91. *Fortsetzung der Actenstücke und Verhandlungen seit dem Jahre*, 2.
92. Oelwein, *Die Regulirung der Donau nächst Linz*, 1–10.
93. Donau-Verein, *Die Thätigkeit des Donau-Vereines*, 3.
94. *Fortsetzung der Actenstücke und Verhandlungen seit dem Jahre 1882*, 2.
95. *Stenographisches Protokoll der Versammlung des Donau-Vereines* (1884), 8–10.
96. *Rechenschaftsbericht... im Jahre 1884*, 166.
97. *Rechenschaftsbericht ... im Jahre 1886*, 86–89; and Oberösterreichischer Landesausschuss, *Bericht über die Thätigkeit ... 1884 ...1890* (1890), 175.
98. *Rechenschaftsbericht... im Jahre 1884*, 69.
99. *Rechenschaftsbericht... im Jahre 1884*, 69–70.
100. *Rechenschaftsbericht... im Jahre 1886*, 86.
101. *Rechenschaftsbericht im Jahre 1886*, 71–72.
102. Judson, *Habsburg Empire*, 246.
103. *Rechenschaftsbericht... im Jahre 1885*, 71–72.
104. Materienbestand Mat. 33, 1889–1925, Donauregulierung I (Umschlagplatz) 310 (alt 188), Archiv der Stadt Linz, Linz, Austria.
105. "Baross ízenete," *Győri Közlöny* (Győr, Hungary), June 21, 1891, p. 4.

106. Bakos, "Győr gyáripara," 45–46.

107. "Baross ízenete," *Győri Közlöny* (Győr, Hungary), June 21, 1891, p. 4.

108. Jankó, *A magyar dunai gőzhajózás*, 91.

109. *Rechenschaftsbericht... im Jahre 1907*, 225.

110. *Rechenschaftsbericht... im Jahre 1909*, 53.

111. Anonymous, "Vízi-Sport," *Vadász- és Versenylap*, May 10, 1883.

112. Anonnymous, "Vízi-Sport," *Vadász- és Versenylap*, June 22, 1882.

113. Anonymous, "Főreáliskolai tanulók-mint mentők," *Pesti Hírlap*, August 10, 1897.

114. Anonymous, "Passautól—Budapestig csónakon," *Pesti Hírlap*, July 20, 1899.

115. Zoltán Pereszlényi, "Fürdő-Élet: Győri fürdőjével," *Pápai Lapok*, July 21, 1895.

116. Lenner, *A győri m. kir. állami főreáliskola*, 246–47.

117. As the project's twenty-four volumes progressed, the state published finished excerpts every two weeks between 1886 and 1903.

118. Habsburg, "Einleitung," 5–6.

119. Exner et al., "Volkswirthschaftliches Leben in Wien," 325.

Conclusion: Collective Action and the Common Good

1. Kożuchowski, *Afterlife of Austria-Hungary*; and Miller and Morelon, *Embers of Empire*.

2. As quoted in Popper, "International Regime," 241.

3. Cioc, *Rhine*, 17.

4. Erste k.k. priv. Donau-Dampfschiffahrts-Gesellschaft, *Geschäfts-Bericht* (1881), 3.

5. Pisecky, "Die Grösste Binnenreederei," 59–60.

6. Extraordinarily, the DDSG's 1867 annual business report failed to mention the Ausgleich among the major factors having an impact on its river traffic that year. Problematic harvests in the rest of Europe caused Romanian grain demand to boom, which benefited the company. However, the DDSG also wrote that poor hydrological conditions adversely affected its navigation. Even as late as 1899, not only did massively poor harvests in the Danube Principalities reduce the number of DDSG grain ships to a tenth of the previous year (1,000 shipments passing the Iron Gates into Hungary in 1898 followed by 97 in 1899), but low water levels in August followed by a once-in-a-century flood in September (described in chapter 4) exacerbated commercial difficulties by halting all steam navigation for several weeks as landing places were rebuilt and destroyed bridge segments were removed

from the river. A dry spell from July to November 1900 effectively halted shipping on the Drave and several other stretches. Erste k.k. priv. Donau-Dampfschiffahrts-Gesellschaft, *Geschäfts-Bericht... 1866 bis... 1867* (1868), 3; Erste k.k. priv. Donau-Dampfschiffahrts-Gesellschaft, *Geschäfts-Bericht ... 1878 bis ... 1879* (1880), 3; Erste k.k. priv. Donau-Dampfschiffahrts-Gesellschaft, *Geschäfts-Bericht . . . Jahr 1900* (1901), 2; and Erste k.k. priv. Donau-Dampfschiffahrts-Gesellschaft, *Geschäfts-Bericht... 1898 bis... 1899* (1900), 1–2.

7. Coates, *Story of Six Rivers*, 23–27.

8. By 1910, over four million already lived in Budapest and Vienna. Four of the empire's twenty largest cities were located on the Danube and nine more were on its tributaries. The four Danube cities in the top twenty were Vienna (1), Budapest (2), Linz (13), and Pressburg/Poszony (16). The nine cities on Danube tributaries were Graz (5) on the Mur, Brünn (8) on the Svratka/Svitava confluence, which flows into the Thaya and later March/Morva River, Szeged (9) on the Tisza, Szabadka (10) on the Tisza, Czernowitz (12) on the Prut, Zagreb (15) on the Sava, Temesvár (17) on the Bega Canal, Nagyvárad/Großwardein (19) on the Sebes-Körös, and Arad (20) on the Maros.

9. Ernest Renan's 1882 lecture "What Is the Nation?" formulated the nation as a product of collective remembering and forgetting and other types of collective experiences that brought people together. He claimed that people's daily actions and decisions reaffirmed a nation's legitimacy to exist as "daily plebiscites."

10. McNeill, *Something New*, 3–17.

11. McNeill and Engelke, *Great Acceleration*.

12. Jing Xuan Teng, "Why Is China Facing Record Floods?" *Phys.org*, July 21, 2021, https://phys.org/news/2021-07-china.html.

13. Jeff Tollefson, "Pivotal Climate Summit Dogged by COVID and Equity Concerns," *Nature*, September 10, 2021, https://www.nature.com/articles/d41586-021-02465-y.

14. Anonymous, "China," *U.S. Energy Information Agency*, September 30, 2020, https://www.eia.gov/international/analysis/country/CHN.

SELECTED BIBLIOGRAPHY

Parliamentary Minutes, Newspaper Articles, and Statistical Reports

Austrian Newspaper Articles: http://anno.onb.ac.at/.
Austrian Parliamentary Discussions, Legal Texts, Imperial Edicts. http://alex.onb.ac.at/.
Hungarian Newspaper Articles. https://adtplus.arcanum.hu/en/.
Hungarian Parliamentary Discussions and Reports. http://www3.arcanum.hu/onap/opt/a110616.htm?v=pdf&a=spec:start.
Hungarian Statistical Reports. *Magyar Statisztikai Évkönyvek* [The Annual Hungarian Statistical Reports] 1870–1914.
Linz Annual Reports. *Rechenschaftsbericht des Gemeinderathes der Landeshauptstadt Linz über seine Thätigkeit, 1878–1914.*
Széchenyi, István. Writings. Magyar Környezetvédelmi és Vízügyi Múzeum [Hungarian Environmental and Hydrological Museum]. Esztergom, Hungary.

Sources

Ágoston, István. *A nemzet inzsellérei: Vízmérnökök élete és munkássága XVIII–XX. sz.* Alsó-Tisza vidéki Vízügyi Igazgatóság, 2001.
Allgemeines Reichs-Gesetz- und Regierungsblatt für das Kaiserthum Oesterreich, Jahrgang 1852. K.k. Hof- und Staatsdruckerei, 1852.
Amts-Blatt der k.k. Bezirkshauptmannschaft St.Pölten: II. Jahrgang. Redaktion und Verlag der k.k. Bezirkshauptmannschaft St. Pölten, 1879.
Andrásfalvy, Bertalan. *A Sárköz és a környező Duna menti területek ősi ártéri gazdálkodása és vízhasználatai a szabályozás előtt.* Ferenc Marczell, 1973.
Andrásfalvy, Bertalan. "Die traditionelle Bewirtschaftung der Überschwem-

Selected Bibliography

mungsgebiete Ungarns (volkstümliche Wassernutzung im Karpatenbecken)." *Acta Ethnographica* 35/1–2 (1989): 39–88.

Anonymous. *Die Schreckens-Tage oder die Überschwemmung in Wien im Jänner 1849*. Josef Stückholzer von Hirschfelt, 1849.

Anonymous. *Die Ueberschwemmung zu Pesth, Ofen und Gran im Monath März 1838 Von einem Augenzeugen*. Wallishaußer, 1838.

Anonymous. *Jubiläums-Ausstellung in Wien 1898: Special-Katalog der Ausstellung der Donau-Regulierungs-Commission in Wien*. Im Verlage der Donau-Regulierungs-Commission, 1898.

Anonymous. *Wien seit 60 Jahren: Zur Erinnerung an die Feier der 60-jährigen Regierung Seiner Majestät des Kaisers Franz Josef I. der Jugend Wiens gewidmet von dem Gemeinderate ihrer Heimatstadt*. Gerlach & Wiedling Buch-und Kunstverlag, 1908.

Arenstein, Joseph. "Die Eisverhältnisse der Donau in Pesth." In Haiginger, *ber die Mitteilungen von Freunden der Naturwissenschaften* 4: 361–63.

Asper, Gottlieb, ed. *Die internationale Fischerei-Conferenz in Wien in der Zeit vom 29. Sept. bis 1. Okt. 1884*. Zürcher und Furrer, 1885.

Az 1892 évi február hó 18-ára hirdetett országyűlés képviselőházának irományai, vol. 29. Pest Könyvnyomda-Részvény-Társaság, 1895.

Az 1892 évi február hó 18-ára hirdetett országyűlés képviselőházának irományai, vol. 31. Pesti Könyvnyomda-Részvény-Társaság, 1895.

Az 1896. évi november hó 23-ára hirdetett országgyűlés képviselőházának naplója, vol. 27. Pesti Könyvnyomda-Részvénytársaság, 1900.

Az 1901. évi október hó 24-ére hirdetett országgyűlés képviselőházának naplója, vol. 1. Pesti Könyvnyomda-Részvénytársaság, 1901.

Az 1901. évi október hó 24-ére hirdetett országgyűlés képviselőházának naplója, vol. 2. Pesti Könyvnyomda-Részvénytársaság, 1902.

Az 1901. évi október hó 24-ére hirdetett országgyűlés képviselőházának naplója, vol. 3. Az Athenaeum Irodalmi és Nyomdai Részvénytársulat, 1902.

Az 1901. évi október hó 24-ére hirdetett országgyűlés képviselőházának naplója, vol. 5. Az Athenaeum Irodalmi és Nyomdai Részvénytársulat, 1902.

Az 1906. évi május hó 19-ére hirdetett országgyűlés képviselőházának naplója, vol. 21. Az Athenaeum Iroldalmi és Nyomdai Részvénytársaság, 1908.

Az 1910. évi június hó 21-ére hirdetett országgyűlés képviselőházának naplója, vol. 17. Az Athenaeum Irodalmi és Nyomdai Részvénytársulat, 1913.

Baader, Joseph von. *Ueber die Verbindung der Donau mit dem Mayn und Rhein und die zweckmäßigste Ausführung derselben*. Seidel, 1822.

Bachinger, Karl. "Das Verkehrswesen." In *Die Hasburgmonarchie, 1848–1914 Band I: Wirtschaftliche Entwicklung*, edited by Alois Brusatti, 278–322. Verlag der Österreichischen Akademie der Wissenschaften, 1973.

Badaeker, Karl. *Southern Germany and Austria, Including Hungary and Transylvania: Handbook for Travelers*. Karl Badaeker, 1883.

Selected Bibliography

Badics, Franz. "Die Bácska." In Vol. 9, *Die österreichisch-ungarische Monarchie in Wort und Bild: Ungarn 2. Band*. k. & k. Hofdruckerei, 1891.

Bakos, Mihály. "Győr gyáripara." In *Győri Szemle*, edited by István Valló, 43–58. Győri Szemle Társaság, 1935.

Ball, Philip. *The Water Kingdom: A Secret History of China*. University of Chicago Press, 2017.

Bałus, Wojciech. *Stadtparks in der österreichischen Monarchie 1765–1918: Studien zur bürgerlichen Entwicklung des urbanen Grüns in Österreich, Ungarn, Kroatien, Slowenien und Krakau aus europäischer Perspektive*. Böhlau Verlag, 2007.

Bana, József, ed. *Győr: a modelváltó város, 1867–1918: Források a dualizmus kori Győr történetéből*. Palatia Nyomda & Kiadó, 2011.

Barany, George. "The Age of Royal Absolutism, 1790–1848." In Sugar, Hanák, and Frank, *History of Hungary*, 174–208.

Barta, Györgyi, Pál Beluszky, Márton Czirfusz, Róbert Győri, György Kukely. *Rehabilitating the Brownfield Zones of Budapest*. Centre for Regional Studies, 2006.

Bastian, Olaf, and Arnd Bernhardt. "Anthropogenic Landscape Changes in Central Europe and the Role of Bioindication." *Landscape Ecology* 8, no. 2 (1993): 139–51.

Bay, Ferenc. *A Győri Lloyd városáért és kereskedeleméért, 1856–1936]*. Baross-Nyomda, 1942.

Beller, Steven. *Francis Joseph*. Longman, 1997.

"Bemerkungen zu dem Gesetzentwurf, betreffend die Beitragsleistung des Staatsschatzes zu den Kosten der Regulirung des Gailflusses in Kärnten." *Stenographische Protokolle über die Sitzungen des Hauses der Abgeordneten des österreichischen Reichsrathes, VIII. Session*. http://alex.onb.ac.at/cgi-content/alex?aid=spa&datum=0008&page=18889&size=45.

Berend, Iván T. "Hungary in the Austro-Hungarian Monarchy and Europe." In *Evolution of the Hungarian Economy 1848—1998. Volume I: One-and-a-Half Centuries of Semi-Successful Modernization 1848—1989*, edited by Iván T. Berend and Tamás Csató, 5–23. Columbia University Press, 2001.

Berend, Iván, and György Ránki. "Economic Problems of the Danube Region After the Break-Up of the Austro-Hungarian." *Journal of Contemporary History* 4, no. 3 (1969): 169–85.

Berry, Helen. "The Great Tyne flood of 1771: Community Responses to an Environmental Crisis in the Early Anthropocene." In *Rivers of the Anthropocene*, edited by Jason Kelly, Philip Scarpino, Helen Berry, James Syvitski, and Michel Meybeck, 119–34. University of California Press, 2018.

Selected Bibliography

Beszédes, József. *Duna-Tiszai hajózható csatornáról: (Über einen schiffbaren Donau-Theiß-Kanal)*. Trattner-Karoly, 1844.

Biró, Marianna, Zsolt Molnár, Dániel Babai et al. "Reviewing Historical Traditional Knowledge for Innovative Conservation Management: A Re-Evaluation of Wetland Grazing." *Science of the Total Environment* 666 (2019): 1114–25.

Bjørnlund, Matthias. "A Fate Worse Than Dying." In *Brutality and Desire: War and Sexuality in Europe's Twentieth Century*, edited by Dagmar Herzog, 16–58. Palgrave, 2009.

Blackbourn, David. *The Conquest of Nature: Water, Landscape, and the Making of Modern Germany*. W. W. Norton, 2006.

Blackbourn, David. "'Time Is a Violent Current': Constructing and Reconstructing German Rivers in Modern German History." In Mauch and Zeller, *Rivers in History*, 11–25.

Blanning, T. C. W. *Joseph II*. Longman, 1995.

Bódi, Laszlo, László Nagy, and Attila Takács. "Review of Historic Floods in Hungary and the Extent of Flooded Areas in Case of Levee Failures." Paper presented at the Sixth Canadian Geohazards Conference, Kingston, Canada, June 15–18, 2014.

Bolzano, Bernard. *Leben Franz Joseph Ritters von Gerstner*. Gottlieb Haase Söhne, 1837.

Borsos, Balázs. "Rivers, Marshes, and Farmlands: Research Perspectives on the Ecological History of Hungary Through Examples of Bodrogköz (NE-Hungary)." *Hungarian Studies* 23, no. 1/2 (2009): 195–210.

Bousfield, Arthur, and Garry Toffoli. *Royal Tours 1786–2010: Home to Canada*. Dundurn Press, 2010.

Boyer, John W. *Political Radicalism in Late Imperial Vienna: Origins of the Christian Social Movement, 1848–1897*. University of Chicago Press, 1981.

Braudel, Fernand. *Civilization and Capitalism, 15th–18th Century, Vol. I: The Structure of Everyday Life, the Limits of the Possible*. University of California Press, 1992.

Brázdil, Rudolf, Zbigniew W. Kundzewicz, Gerardo Benito, Gaston Demarée, Neil Macdonald, and Lars A. Roald. "Historical Floods in Europe in the Past Millennium." In *Changes in Flood Risk in Europe*, edited by Zbigniew W. Kundzewicz, 121–66. CRC Press, 2012.

Bright, Richard. *Travels from Vienna Through Lower Hungary*. Archibald Constable and Company, 1818.

Brönnimann, Stefan, Sam White, and Victoria Slonosky. "Climate from 1800 to 1970 in North America and Europe." In White, Pfister, and Mauelshagen, *Palgrave Handbook of Climate History*, 309–20.

Brunner, Karl, and Petra Schneider. *Umwelt Stadt: Geschichte des Natur- und Lebensraumes Wien*. Böhlau Verlag, 2005.

Selected Bibliography

Buffe, Noël. *Marines du Danube, 1526–1918*. Lavauzelle, 2011.

Cioc, Mark. *The Rhine: An Eco-Biography, 1815–2000*. University of Washington Press, 2002.

Coates, Peter. *A Story of Six Rivers: History, Culture and Ecology*. Reaktion Books, 2013.

Coen, Deborah R. "Climate and Circulation in Imperial Austria." *Journal of Modern History* 82, no. 4 (2010): 839–75.

Coen, Deborah R. *Climate in Motion: Science, Empire, and the Problem of Scale*. University of Chicago Press, 2018.

Cohen, Gary. "Cultural Crossings in Prague, 1900: Scenes from Late Imperial Austria." *Austrian History Yearbook* 45 (2014): 1–30.

Cohen, Gary B. Cohen. "Nationalist Politics and the Dynamics of State and Civil Society in the Habsburg Monarchy, 1867–1914," *Central European History* 40 (2007): 241–78.

Collins, Timothy M., Edward K. Muller, and Joel A. Tarr. "Pittsburgh's Three Rivers: From Industrial Infrastructure to Environmental Asset." In Mauch and Zeller, *Rivers in History*, 41–62.

Csendes, Péter, and András Sipos, eds. *Budapest und Wien: Technischer Fortschritt und Urbaner Aufschwung im 19. Jahrhundert*. Deuticke, 2003.

Csikvári, Jákó. *A közlekedési eszközök: A vasutak, posták, távirdák és a gőzhajózás története*. Vol 2. Franklin-Társulat Könyvnyomdája, 1883.

Csorba, László. "Transition from Pest-Buda to Budapest, 1815–1873." In *Budapest: A History from Its Beginnings to 1998*, edited by András Gerő and János Poór, 69–101. Atlantic Research and Publications, 1997.

Csukovits, Anita, and Katalin Forró, eds. *Duna: Az ember és a folyó*. Pest Megyei Múzeumok Igazgatósága, 2008.

Czeike, Felix. "Wiener Neustädter Kanal." *Historisches Lexikon Wien*, s.v., 638–39. Kremayr & Scheriau, 2004.

Czoernig, Carl Freiherr von. *Österreich's Neugestaltung, 1848–58*. J. G. Cotta'scher Verlage, 1858.

Deák, István. *Beyond Nationalism: A Social and Political History of the Habsburg Officer Corps, 1848–1918*. Oxford University Press, 1990.

Deak, John. *Forging a Multinational State: State Making in Imperial Austria from the Enlightenment to the First World War*. Stanford University Press, 2015.

Demény, Zsuzsa, and Gyula Holka, eds. *Statistics of the Centuries*. Hungarian Central Statistical Office, 2002.

Dénes, Ihrig, ed. *A magyar vízszabályozás története*. Budapest, 1973.

Denkschrift über das Projekt eines Donau-Save-Schiffahrt-Canales zwischen Vukovár und Samač in Slavonien und der Militärgränze. Gerold Sohn, 1869.

Dezsényi, Miklós, and Ferenc Hernády. *A Magyar Hajózás Története*. Műszaki Könyvkiadó, 1967.

Selected Bibliography

Dick, Thomas. *The Christian Philosopher or, the Connection of Science and Philosophy with Religion.* 3rd ed. Chalmers and Collins, 1825.

"Die älteste Versicherung Tirols und ihre Vorgeschichte." *Tirol.gv.at.* https://www.tirol.gv.at/fileadmin/themen/kunst-kultur/landesarchiv/downloads/versicherung.PDF.

"Die Au nach dem Hochwasser." *Nationalpark Donau Auen.* https://www.donauauen.at/wissen/natur-wissenschaft/die-donau/die-au-nach-dem-hochwasser.

Die Donau und ihr Einzugsgebiet: Eine hydrologische Monographie, Teil 1 Texte. Regionale Zusammenarbeit der Donauländer, 1986.Dóka, Klára. *A vízimunkálatok irányítása és jelentősége az ország gazdasági életében, 1772–1918.* Mezőgazdasági Ügyvitelszervezési Iroda, 1987.

Dóka, Klára. "A vízügyi szakigazgatás fejlődése, I. rész (1772–1867)." *Vízügyi Közlemények* 4 (1982): 515–29.

Dóka, Klára. "A vízügyi szakigazgatás fejlődése, II. rész (1867–1948)." *Vízügyi Közlemények* 1 (1983): 54–68.

Domaszewski, Viktor von. *Betrachtungen über die Ausnützung des Marchfeldes.* Vienna, 1878.

Domaszewski, Viktor von. *Das Wasser als Quelle der Verwüstungen und des Reichthums: Nach der Natur geschildert.* Vienna, 1879

Domaszewski, Viktor von. *Nutzwasser- und Wasserkraftfrage der Stadt Wien.* Vienna, 1878.

Domaszewski, Viktor von. *Regulirung des Wienflusses.* Vienna, 1878.

Domaszewski, Viktor von. *Was kostet der unvermeidliche Wasserkrieg im österr.-ungar: Donau-Gebiete?* Vienna, 1879.

Donau-Regulirungs-Commission. *Beschreibung der Arbeiten der Donau-Regulirung bei Wien: herausgegeben aus Anlaß der feierlichen Eröffnung der Schiffahrt im neuen Strombette am 30. Mai 1875.* k. k. Hof- u. Staatsdruckerei, 1875.

Donau-Verein. *Die Thätigkeit des Donau-Vereines im ersten Jahrzehnte seines Bestandes, 1879–88.* Verlag des Donau-Vereines, 1889.

Dorner, Joseph von. "Die neue Strasse durch die Kliffura, Engpass im walachisch-banatischen Grenz-Gebiet." In *Panorama der Österreichischen Monarchie oder malerisch-romantisches Denkbuch der schönsten und merkwürdigsten Gegenden derselben, der Gletscher, Hochgebirge, Alpenseen und Wasserfälle, bedeutender Städte mit ihren Kathedralen, Pallästen und alterthümlichen Bauwerken, berühmter Badeörter, Schlösser, Burgen und Ruinen, sowie der interessantesten Donau-Ansichten mit Stahlstichen von vorzüglichsten englischen und deutschen Künstlern nach eigends zu diesem Werke aufgenommenen Originalzeichnungen,* 3:11–15. C. A. Hartleben's Verlag, 1840.

Doyle, Martin. *The Source: How Rivers Made America and America Remade Its Rivers.* W. W. Norton, 2018.

Selected Bibliography

Dürrnberger, Adolf. "Das Donauthal von Passau bis Linz." In Vol. 6, *Die österreichisch-ungarische Monarchie in Wort und Bild: Oberösterreich und Salzburg*. k. & k. Hofdruckerei, 1889.

Eder, Ernst Gerhard. *Bade- und Schwimmkultur in Wien*. Böhlau, 1995.

The Edinburgh Encyclopedia. William Blackwood and John Waugh, 1830.

Edvinsson, Rodney. "Historical Currency Converter (Test Version 1.0)." http://www.historicalstatistics.org/Currencyconverter.html.

Ekmekcioglu, Lerna. "A Climate for Abduction, a Climate for Redemption: The Politics of Inclusion During and After the Armenian Genocide." *Comparative Studies in Society and History* 55 (2013): 522–53.

Ember, Győző. "A magyarországi építészeti igazgatóság történetének vázlata." *Levéltári Közlemények A Magyar Országos Levéltár Folyóirata* 20–23 (1945): 345–75.

"Ermächtigung der an der Donau gelegenen österr. Gränzzollämter, das auf Schiffen der ößterr. Donau-Dampfschiffahrtsgesellschaft in der Einfuhr vorkommende Getreide, dann Knoppern, Summach u. dgl. ohne Erklärung und Gewichts-Constatirung an ein Amt im Innern anzuweisen." *Verordnungs-Blatt für den Dienstbereich österreichischen Finanzministeriums, Jahrgang 1854*. K.k. Hof- und Staatsdruckerei, 1854.

Erste k.k. priv. Donau-Dampfschiffahrts-Gesellschaft. *Geschäfts-Bericht der Betriebs-Direction über das Verwaltungsjahr vom 1. December 1866 bis 30. November 1867*. Selbstverlag der Gesellschaft, 1868.

Erste k.k. priv. Donau-Dampfschiffahrts-Gesellschaft. *Geschäfts-Bericht der Betriebs-Direction über das Verwaltungsjahr vom 1. December 1878 bis 30. November 1879*. Selbstverlag der Gesellschaft, 1880.

Erste k.k. priv. Donau-Dampfschiffahrts-Gesellschaft. *Geschäfts-Bericht der Betriebs-Direction über das Verwaltungsjahr vom 1. December 1879 bis 30. November 1880*. Selbstverlag der Gesellschaft, 1881.

Erste k.k. priv. Donau-Dampfschiffahrts-Gesellschaft. *Geschäftsbericht der Betriebs-Direction über das Verwaltungsjahr vom 1. December 1887 bis 30. November 1888*. Carl Gerold's Son, 1889.

Erste k.k. priv. Donau-Dampfschiffahrts-Gesellschaft. *Geschäfts-Bericht der Betriebs-Direction über das Verwaltungsjahr vom 1. December 1898 bis 30. November 1899*. Selbstverlag der Gesellschaft, 1900.

Erste k.k. priv. Donau-Dampfschiffahrts-Gesellschaft. *Geschäfts-Bericht für das Jahr 1896*. Selbstverlag der Gesellschaft, 1897.

Erste k.k. priv. Donau-Dampfschiffahrts-Gesellschaft. *Geschäfts-Bericht für das Jahr 1900*. Selbstverlag der Gesellschaft, 1901.

"Erzsébet királyné szobra." *BudapestCity.org*. https://budapestcity.org/egyeb/tervek/Erzsebet-kiralyne-szobra/index.html.

Exner, Wilhelm Franz Rudolf von Grimburg, Adolf von Guttenberg, W. Hecke, Emanuel Sar. "Volkswirthschaftliches Leben in Wien." In Vol. 1, *Die*

Selected Bibliography

österreichisch-ungarische Monarchie in Wort und Bild: Wien und Niederösterreich. k. & k. Hofdruckerei, 1886.

Fagan, Brian. *The Little Ice Age: How Climate Made History 1300–1850*. Basic Books, 2002.

Fallon, John. "Passau to Pesth, along the Danube. Part II." *Irish Monthly* 12, no. 135 (1884): 447–54.

Fassmann, Heinz. "City-Size Distribution in the Austrian-Hungarian Monarchy, 1857–1910: A Rank-Size Approach." *Historical Social Research / Historische Sozialforschung*, no. 38 (April 1986): 3–24.

Fejér, László. *Árvizek és belvizek szorításában: A vízkárelhárítás jogi szabályozásának fejlődése, különös tekintettel a védekezés szervezeti oldalára és gazdasági feltételeire*. Vízügyi Történeti Füzetek, 1997.

Fejér, László, ed. *Vizeink krónikája*. Vízügyi Múzeum, Levéltár és Könyvgyűjtemény, 2001.

Földes, Béla, ed. *Nemzetgazdasági Statisztikai Évkönyv*. A Magyar Tud. Akadémia Könyvkiadó-Hivatala, 1883.

Forgách, Ludwig von. *Die schiffbare Donau von Ulm bis in das Schwarze Meer: den Mitgliedern des verfaßunggebenden Reichstages zur gütigen Einsicht*. August Osterrieth, 1848.

Forth, Christopher E. *The Dreyfus Affair and the Crisis of French Manhood*. Johns Hopkins University Press, 2006.

Fortsetzung der Actenstücke und Verhandlungen seit dem Jahre 1882 bezugnehmend auf die Versandung des Donau-Landungsplatzes und des sogenannten Fabriksarmes in Linz. Linz: Verlag der Gemeinde Linz, 1883.

Fray, Franz B. *Allgemeiner Handlungs-Gremial-Almanach für den oesterreichischen Kaiserstaat*. Vienna, 1837.

Fritsch, Anton. "Eine Reise nach dem Banate." In *Erinnerungsschrift zum Gedächtnisse an die Jahresversammlung der deutschen Ornithologen-Gesellschaft abgehalten in Halberstadt vom 11. Bis 14. Juli 1853*, edited by Jean Cabanis, 33–38. Cassel: Verlag von Theodor Fischer, 1853.

Gatejel, Luminita. *Engineering the Lower Danube: Technology and International Cooperation in an Imperial Borderland*. CEU Press, 2022.

Géra, Eleonóra. "Wechselwirkung zwischen Donau und Alltag in Ofen-Pest in den Jahren 1686 bis 1800." In Tamáska and Szabó, *Donau-Stadt-Landschaften*, 59–71.

Geschäfts-Bericht der Betriebs-Direction der ersten k.k. priv. Donau-Dampfschiffahrt-Gesellschaft über das Verwaltungsjahr vom 1. December 1857 bis 30. November 1858. Carl Gerold's Sohn, 1859.

Geschäfts-Bericht für das Jahr 1896. Selbstverlag der Gesellschaft, 1897.

"Geschichte." *Zentralanstalt für Meteorologie und Geodynamik*. http://www.zamg.ac.at/cms/de/topmenu/ueber-uns/geschichte.

Gierlinger, Sylvia, Gertrud Haidvogl, Simone Gingrich, and Fridolin Krausmann. "Feeding and Cleaning the City: The Role of the Urban Waterscape in Provision and Disposal in Vienna During the Industrial Transformation." *Water History* 5 (2013): 219–39.

Gingrich, Simone, Gertrud Haidvogl, and Fridolin Krausmann. "The Danube and Vienna: Urban Resource Use, Transport and Land Use 1800 to 1910." *History of Urban Environmental Imprint* 12, no. 2 (2012): 283–94.

Giosan, Liviu, Marco J. L. Coolen, Jed O. Kaplan et al. "Early Anthropogenic Transformation of the Danube-Black Sea System." *Scientific Reports* 2, no. 582 (2012): 1–6.

Glaser, Rüdiger, Dirk Riemann, Johannes Schönbein et al. "The Variability of European Floods Since AD 1500." *Climatic Change* 101 (2010): 235–56.

Glassl, Horst. "Der Ausbau der ungarischen Wasserstraßen in den letzten Regierungsjahren Maria Theresias." *Ungarn-Jahrbuch: Zeitschrift für die Kunde Ungarns und verwandte Gebiete* 2 (V. Hase & Koehler Verlag, 1970): 34–66.

Gonda, Béla. "A magyar Duna." In Vol. 16, in *Az Osztrák-Magyar Monarchia írásban és képben Magyarország IV*. Magyar Királyi Államnyomda, 1896.

Gonda, Béla. *Az Al-Dunai Vaskapu és az Ottani Többi Zuhatag Szabályozása*. Országgyűlési Értesítő Kő- és Könyvnyomda Részvénytársaság, 1896.

Gonda, Béla. *Die ungarische Schiffahrt*. Technisch-Litterarische und Druckerei-Unternehmung, 1899.

Good, David F. *The Economic Rise of the Habsburg Empire, 1750–1914*. University of California Press, 1984.

Göőz, József. *Budapest története*. Róbert Lampel, 1896.

Gottlieb, Asper, ed. *Die internationale Fischerei-Conferenz in Wien in der Zeit vom 29. Sept. bis 1. Okt. 1884*. Zürcher und Furrer, 1885.

Gráfik, Imre. *Hajózás és gabonakereskedelem: "Gabonakonjunktúra vízen,"* Pro Pannónia Kiadói Alapítvány, 2004.

Grever, Maria. "Staging Modern Monarchs: Royalty at the World Exhibitions of 1851 and 1867." In *Mystifying the Monarch: Studies on Discourse, Power, and History*, edited by Jeroen Deploige and Gita Deneckere, 161–80. Amsterdam University Press, 2006.

Grössing, Funk, Sauer, and Binder. *Rot-Weiss-Rot auf blauen Wellen 150 Jahre DDSG*. Eigenverlag, 1979.

Gruber, Helmut. *Red Vienna: Experiment in Working-Class Culture, 1919–1934*. Oxford University Press, 1991.

Grübler, Arnulf. *The Rise and Fall of Infrastructures: Dynamics of Evolution and Technological Change in Transport*. Physica-Verlag, 1990.

Gruffat, Corentin. "The Beautiful Public Danube: Water Uses, Water Rights, and the Habsburg Imperial State in the Mid-Nineteenth Century." *Austrian History Yearbook* 54 (2023): 136–58.

Gunst, Péter. *Agrarian Development and Social Change in Eastern Europe, 14th–19th Centuries*. Variorum, 1996.

Selected Bibliography

Gyáni, Gábor. "Das Verhältnis von Urbanisation, Großstadtentwicklung und der Donau in Budapest des 19. Und 20. Jahrhunderts." In Tamáska and Szabó, *Donau-Stadt-Landschaften*, 73–85.

Habsburg, Crown Prince Rudolf von. "Einleitung." In Vol. 2, *Die österreichisch-ungarische Monarchie in Wort und Bild: Übersichtsband Naturgeschichtlicher Theil*. k. & k. Hofdruckerei, 1887.

Habsburg, Rudolf, E. F. von Homeyer, and A. Brehm. "Zwölf Frühlingstage an der mittleren Donau." *Journal für Ornithologie* 27, no. 145 (1879): 1–83.

Hagen, Anna, and Friedrich Hauer. "Hygiene und Wasser in der städtebaulichen Fachliteratur um 1900." *Materialien zur Umweltgeschichte Österreichs* 7 (Zentrum für Umweltgeschichte, 2015): 1–58.

Haidinger, Wilhelm, ed. *Bericht uber die Mitteilungen von Freunden der Naturwissenschaften in Wien*, vol. 4. Wilhelm Braumüller, 1848.

Haidvogl, Gertrud, and Simone Gingrich. "Wasserstrasse Donau: Transport und Hnadel im Machland und auf der Donau im 19. Und 20. Jahrhundert." In Winiwarter and Schmid, *Umwelt Donau*, 91–103.

Haidvogl, Gertrud, Friedrich Hauer, Severin Hohensinner et al. *Wasser, Stadt, Wien: Eine Umweltgeschichte*. Center for Environmental History, 2020.

Hajnal, Henry. *The Danube: Its Historical, Political and Economic Importance*. Martinus Nijhoff, 1920.

Hakkarainen, Heidi. "City Upside Down: Laughing at the Flooding of the Danube in Late Nineteenth-Century Vienna." In *Catastrophe, Gender and Urban Experience, 1648–1920*, edited by Deborah Simonton and Hannu Salmi, 157–76. Routledge, 2017.

Hakkarainen, Heidi. *Comical Modernity: Popular Humour and the Transformation of Urban Space in Late Nineteenth-Century Vienna*. Berghahn Books, 2019.

Halter, Rudolf. "Das zentraleuropäisclie Wasserstraßennetz: Vortrag, gehalten den 28. November 1917." *ZOBODAT Zoological Botanical Database*. http://www.zobodat.at/pdf/SVVNWK_58_0063-0090.pdf.

Hamann, Brigitte. "Empress Elisabeth, 1837–1898." In *The Imperial Style: Fashions of the Hapsburg Era*, edited by Polly Cone, 129–52. Metropolitan Museum of Art, 1980.

Hammerl, Christa. "Viennese School of Climatology." In *Oxford Research Encyclopedia*. Oxford University Press, 2018. DOI: 10.1093/acrefore/9780190228620.013.701.

A Handbook for Travelers in Southern Germany; Being a Guide to Bavaria, Austria, Tyrol, Salzburg, Styria, & the Austrian and Bavarian Alps, and the Danube from Ulm to the Black Sea. John Murray and Sons, 1837.

Hartwell, Edward Mussey. "Public Baths in Europe." In *Bulletin of the De-*

Selected Bibliography

partment of Labor, edited by Carroll D. Wright, 434–86. Government Printing Office, 1897.

Haselsteiner, Horst. "Cooperation and Confrontation Between Rulers and the Noble Estates, 1711–1790." In Sugar, Hanák, and Frank, *History of Hungary*, 149–54.

Hauer, Friedrich. "Wien und die Donau(auen): Zur Entstehung einer Stadtlandlandschaft." In Tamáska and Szabó, *Donau-Stadt-Landschaften*, 121–33.

Hauer, Friedrich, Severin Hohensinner, and Christina Spitzbart-Glasl. "How Water and Its Use Shaped the Spatial Development of Vienna." *Water History* 8 (2016): 301–28.

Hauer, Friedrich, and Christina Spitzbart-Glasl. "Nebenvorteile und Erbschaften einer Wasserstraße, Bedeutung und Permanenz von sekundären Nutzungen am Wiener Neustädter Kanal in Wien." *Wiener Geschichtsblätter 72. Jahrgang* 2 (2017): 1–33.

Hegyfoky, J. "Wasserstand der Flüsse und Niederschlag in Ungarn." *Mathematische und Naturwissenschaftliche Berichte aus Ungarn* 27, no. 1 (1897): 239–84.

Heksch, Alexander F. *Illustrirter Führer auf der Donau von Regensburg bis Sulina*. A. Hartleben's Verlag, 1880.

Heliser, Jósef. *Rövid tudósítás az 1838-iki esztergomi árvízről, annak következményeiről, a kárvallottak számára béfolytt segedelmekről, és ezeknek felosztásukról*. Esztergomi k. Beimel J. betűivel, 1839.

Herold, Roland P. *Brigittenau: Von der Au zum Wohnbezirk*. Mohl Verlag, 1992.

Hieronymi, Károly. *A budapesti Duna-szakasz szabályozása*. Pesti Könyvnyomda-Részvény Társaság Kiadványa, 1880.

Hieronymi, Károly. *A közlekedés*. Pest, 1869.

Hieronymi, Károly. *A közmunkaügyek állami kezelése. I. Rész: A közmunkaügyek állami kezelése Francziaországban*. Budapest, 1874.

Hieronymi, Károly. *A közutak föntartásáról: Az útfentartás különböző módjainak, a fentartási költség tényezőinek és ezek megszerzésének ismertetése. Főleg franczia kutfők után*. Pest, 1868.

Hieronymi, Károly. *Die Theissregulirung*. Budapest, 1888.

Hinkel, Raimund. *Wien an der Donau: Der große Strom, seine Beziehung zur Stadt und die Entwicklung der Schiffahrt im Wandel der Zeiten*. Christian Brandstätter, 1995.

Hoffman, Alfred. "Die Donau und Österreich." *Südosteuropa-Jahrbuch* 5 (Südosteuropa-Verlagsgesellschaft m.b.H., 1961): 28–42.

Hofman, F. *Die Ueberschwemmung von Wien und seiner Umgebung im Februar 1862*. Alexander Eurich, 1862.

Selected Bibliography

Hohensinner, Severin. "Die Verwandlung der Donau—Eine kaum zu bändigende Flusslandschaft." In *Die Donau—Eine Reise in die Vergangenheit. Katalog zur gleichnamigen Ausstellung der Österreichischen Nationalbibliothek 29. April–7. Nov. 2021*, 29–39. Kremayr & Scheriau, 2021.

Hohensinner, Severin. "Historische Hochwässer der Wiener Donau und ihrer Zubringer." *Materialien zur Umweltgeschichte Österreichs* 1 (Zentrum für Umweltgeschichte, 2015): 1–51.

Hohensinner, Severin. "'Sobald jedoch der Strom einen anderen Lauf nimmt ...': Der Wandel der Donau vom 18. Bis 20. Jahrhundert." In Winiwarter and Schmid, *Umwelt Donau*, 39–55.

Hohensinner, Severin. "'Wie viele Fahrzeuge liegen in den Schottermassen begraben?' Die Schifffahrt auf der unregulierten Donau." In Winiwarter and Schmid, *Umwelt Donau*, 105–17.

Hohensinner, Severin, Anton Drescher, Otto Eckmüllner et al. *Genug Holz für Stadt und Fluss? Wiens Holzressourcen in dynamischen Donau-Auen*. Verlag Guthmann-Peterson, 2016.

Hohensinner, Severin, and Andreas Hahmann. "Historische Wasserbauten an der Wiener Donau und ihren Zubringern." *Materialien zur Umweltgeschichte Österreichs* 2 (Zentrum für Umweltgeschichte, 2015): 1–351.

Hohensinner, Severin, and Gertrud Haidvogl. "Zu viel Wasser: Hochwassergefahr und Praktiken des Überschwemmungsschutzes." In *Wasser, Stadt, Wien: Eine Umweltgeschichte*, edited by the Center for Environmental History, 160–71. University of Natural Resources and Life Sciences (BOKU), 2019.

Hohensinner, Severin, and Friedrich Hauer. "Durchstich, Kai und Häusergerümpel—Die Donauregulierung 1870–1876 als landschafts- und städtebauliches Großprojekt." *Studien zur Wiener Geschichte—Jahrbuch des Vereins für Geschichte der Stadt Wien* 79 (2023): 171–221.

Hohensinner, Severin, Mathew Herrnegger, Alfred P. Blaschke et al. "Type-Specific Reference Conditions of Fluvial Landscapes: A Search in the Past by 3D-Reconstruction." *Catena* 75 (2008): 200–215.

Hohensinner, Severin, Matthias Jungwirth, S. Muhar, and S. Schmutz. "Spatio-Temporal Habitat Dynamics in a Changing Danube River Landscape 1812–2006." *River Research and Applications* 27 (2011): 939–55.

Hohensinner, Severin, Bernhard Lager, Christoph Sonnlechner et al. "Changes in Water and Land: The Reconstructed Viennese Riverscape from 1500 to the Present." *Water History* 5, no. 2 (2013): 145–72.

Hohensinner, Severin, and Martin Schmid. "The More Dikes, the Higher the Floods: Vienna and its Danube Floods." In Tamáska and Szabó, *Donau-Stadt-Landschaften*, 211–27.

Hohensinner, Severin, Christoph Sonnlechner, Martin Schmid, and Verena Winiwarter. "Two Steps Back, One Step Forward: Reconstructing the

Dynamic Danube Riverscape Under Human Influence in Vienna." *Water History* 5 (2013):121–43.

Hull, Isabel V. *Absolute Destruction: Military Culture and the Practices of War in Imperial Germany.* Cornell University Press, 2004.

"Hungary." In *Supplement to the Fourth, Fifth, and Sixth Editions of the Encyclopedia Britannica,* vol. 5. Archibald Constable and Company, 1824.

Husain, Faisal H. *Rivers of the Sultan: The Tigris and Euphrates in the Ottoman Empire.* Oxford University Press, 2021.

Huszár, Zoltán. "Zu den Beziehungen zwischen Pécs und der Ersten Donau-Dampfschiffahrts Gesellschaft (DDSG), mit besonderer Berücksichtigung der Sozialpolitik." In *Donau-Schiffahrt,* edited by Arbeitskreis Schiffahrtsmuseum Regensburg, e.V., 137–46. Selbstverlag, 2004.

Ingrao, Charles W. *The Habsburg Monarchy, 1618–1815.* 2nd ed. Cambridge University Press, 2000.

Jäckel, Andreas Johann. "Der Vögelzug und anderweitige Wahrnehmungen über die Vogelwelt Bayerns, im Jahre 1853/54." *Journal für Ornithologie* 2, no. 12 (1854): 481–502.

Jankó, Béla. *A Magyar dunai gőzhajózás története.* Gépipari tudományos egyesület, 1968.

Jászi, Oszkár. *The Dissolution of the Habsburg Monarchy.* University of Chicago, 1929.

Jekelfalussy, József. *The Millennium of Hungary and Its People.* Pesti Könyvnyomda-Részvénytársaság, 1897.

Jekelfalussy, József, and Gyula Vargha, eds. *Közgazdasági és statisztikai évkönyv, újabb ötödik évfolyam, 1891.* Pesti Könyvnyomda-Részvény-Társaság, 1891.

Jobst, Clemens, and Helmut Stix. "Florin, Crown, Schilling and Euro: An Overview of 200 Years of Cash in Austria." *Monetary Policy & the Economy* (2016): 94–119.

Jókai, Mór. "Bevezetés." In Vol. 12, *Az Osztrák-Magyar Monarchia írásban és képben: Magyarország III (I).* Magyar Királyi Államnyomda, 1893.

Jókai, Mór. *Timar's Two Worlds.* Translated by Hegan Kennard. D. Appleton and Company, 1895.

Judson, Pieter M. *Guardians of the Nation: Activists on the Language Frontiers of Imperial Austria.* Harvard University Press, 2006.

Judson, Pieter M. *The Habsburg Empire: A New History.* Harvard University Press, 2016.

Juhász, Endre. *A csatornázás története.* T-Mart Press Kiadó és Nyomdaipari Kft, 2008.

Jungwirth, Matthias, Gertrud Haidvogl, Severin Hohensinner, Herwig Waidbacher, and Gerald Zauner. *Österreichs Donau: Landschaft-Fisch-Geschichte.* Institut für Hydrobiologie und Gewässermanagement (BOKU), 2014.

Selected Bibliography

Kállay, István. "Ungarischer Donauhandel, 1686–1848." *Historisches Jahrbuch der Stadt Linz, 1987* (Archiv der Stadt Linz, 1988): 41–49.

Károlyi, Zsigmond. "A magyar vízi munkálatok rövid története különös tekintettel a vizek szabályozására." In *A magyar vízszabályozás története*, edited by Ihrig Dénes, 23–147. Országos Vízügyi Hivatal, 1973.

Katalin, Péter. "The Later Ottoman Period and Royal Hungary, 1606–1711." In *A History of Hungary*, edited by Peter F. Sugar, Péter Hanák, and Tibor Frank, 100–120. Indiana University Press, 1994.

King, Jeremy. *Budweisers into Germans and Czechs: A Local History of Bohemian Politics, 1848–1948*. Princeton University Press, 2002.

K.k. Direction der administrativen Statistik, ed. "Handel des österreichischen Zollgebiets." In *Tafeln zur Statistik der österreichischen Monarchie für die Jahre 1845 und 1846*, vol. 2. K.k. Hof und Staatsdruckerei, 1851.

K.k. Direction der administrativen Statistik, ed. *Tafeln zur Statistik der österreichischen Monarchie für das Jahr 1841, 14. Jahrgang*. K.k. Hof- und Staats-Druckerei, 1844.

K.k. Direction der administrativen Statistik, ed. *Tafeln zur Statistik der österreichischen Monarchie für das Jahr 1842, 15. Jahrgang*. K.k. Hof- und Staats-Druckerei, 1846.

K.k. Direction der administrativen Statistik, ed. *Tafeln zur Statistik der österreichischen Monarchie für das Jahr 1843, 16. Jahrgang*. K.k. Hof- und Staats-Druckerei, 1847.

K.k. Direction der administrativen Statistik, ed. *Tafeln zur Statistik der österreichischen Monarchie für das Jahr 1844, 17. Jahrgang*. K.k. Hof- und Staats-Druckerei, 1847.

K.k. Direction der administrativen Statistik, ed. *Tafeln zur Statistik der österreichischen Monarchie für die Jahre 1847 und 1848, Erster Theil*. K.k. Hof- und Staats-Druckerei, 1853.

K.k. Direction der administrativen Statistik, ed. *Tafeln zur Statistik der österreichischen Monarchie für die Jahre 1849–51*. K.k. Hof- und Staatsdruckerei, 1856.

K.k. Direction der administrativen Statistik, ed. *Tafeln zur Statistik der österreichischen Monarchie Neue Folge I. Das Jahr 1851 mit übersichtlicher Einbeziehung der Jahre 1849 und 1850*. K.k. Hof- und Staatsdruckerei, 1856.

K.k. Direction der administrativen Statistik, ed. *Tafeln zur Statistik der österreichischen Monarchie für die Jahre 1852–54*. K.k. Hof- und Staatsdruckerei, 1859.

K.k. Direction der administrativen Statistik, ed. *Tafeln zur Statistik der österreichischen Monarchie für die Jahre 1855–57*. K.k. Hof- und Staatsdruckerei, 1861.

K.k. Direction der administrativen Statistik, ed. *Tafeln zur Statistik der öster-

Selected Bibliography

reichischen Monarchie für die Jahre 1858–59. K.k. Hof- und Staatsdruckerei, 1862.

K.k. Statistische Central-Commission, ed. "Waarenverkehr zwischen Oesterreich und Ungarn in den Jahren 1884 bis 1891." *Oesterreichische Statistik* 37, no. 4/2 (K.k. Hof- und Staatsdruckerei, 1894): 158–69.

K.k. Statistische Central-Commission, ed. *Oesterreichisches Statistisches Handbuch für die im Reichsrathe vertretenen Königreiche und Länder nebst einem Anhange für die Gemeinsamen Angelegenheiten der österreichisch-ungarischen Monarchie. Sechster Jahrgang 1887.* K.k. Statistische Central-Commission, 1888.

K.k. Statistische Central-Commission, ed. *Österreichisches Statistisches Handbuch für die im Reichsrathe Vertretenen Königreiche und Länder nebst einem Anhange für die gemeinsamen Angelegenheiten der österreichisch-ungarischen Monarchie, 1897.* K.k. Statistische Central-Commission, 1898.

K.k. Statistische Central-Commission, ed. *Österreichisches Statistisches Handbuch für die im Reichsrathe Vertretenen Königreiche und Länder nebst einem Anhange für die gemeinsamen Angelegenheiten der österreichisch-ungarischen Monarchie, 1907.* K.k. Statistische Central-Commission, 1908.

K.k. Statistische Central-Commission, ed. *Österreichisches Statistisches Handbuch für die im Reichsrathe Vertretenen Königreiche und Länder nebst einem Anhange für die gemeinsamen Angelegenheiten der österreichisch-ungarischen Monarchie, 1914.* K.k. Statistische Central-Commission, 1916.

K.k. Statistische Central-Commission, ed. *Statistik des Verkehres in den im Reichsrathe vertretenen Königreichen und Ländern vornehmlich für die Jahre 1881 bis 1891.* K.k. Hof- und Staatsdruckerei, 1893.

K.k. Statistische Central-Commission, ed. *Statistik des Verkehrs in den im Reichsrathe vertretenen Königreichen und Ländern vornehmlich für die Jahre 1894 und 1895.* K.k. Hof- und Staatsdruckerei, 1897.

K.k. Statistische Central-Commission, ed. *Statistische Monatschrift: XII. Jahrgang.* K.k. Hof- und Universitäts-Buchhändler, 1886.

K.k. Statistische Central-Commission, ed. *Statistisches Jahrbuch für das Jahr 1868.* k.k. Hof- und Staatsdruckerei, 1870.

K.k. Statistische Central-Commission, ed. *Statistisches Jahrbuch für das Jahr 1869.* Carl Gerold's Son, 1871.

K.k. Statistische Central-Commission, ed. *Statistisches Jahrbuch für das Jahr 1869.* K.k. Hof- und Staatsdruckerei, 1871.

K.k. Statistische Central-Commission, ed. *Statistisches Jahrbuch für das Jahr 1870.* k.k. Hof- und Staatsdruckerei, 1872.

K.k. Statistische Central-Commission, ed. *Statistisches Jahrbuch für das Jahr 1877*, vol. 4. K.k. Hof- und Staatsdruckerei, 1880.

Selected Bibliography

K.k. Statistische Central-Commission, ed. *Tafeln zur Statistik der österreichischen Monarchie, Die Jahre 1860 bis 1865 umfassend*. K.k. Hof- und Staatsdruckerei, 1871.

K.k. Statistische Zentral-Commission, ed. "Der Zwischenverkehr der im Reichsrate vertretenen Königreiche und Länder mit den Ländern der ungarischen Krone für die Jahren 1902 bis 1905," *Oesterreichische Statistik* 82, no. 3/2 (Hof- und Staatsdruckerei, 1905): lvii–lxxii.

K.k. Statistische Zentral-Commission, ed. "Der Zwischenverkehr der im Reichsrate vertretenen Königreiche und Länder mit den Ländern der ungarischen Krone in den Jahren 1900 und 1901." *Oesterreichische Statistik* 68, no. 1/2 (Hof- und Staatsdruckerei, 1905): lxiii–lxxi.

K.k. Statistische Zentralkommission, ed. "Der Zwischenverkehr der im Reichsrate vertretenen Königreiche und Länder mit den Ländern der ungarischen Krone in den Jahren 1906 u. 1907." *Oesterreichische Statistik* 91, no. 4/2 (Hof- und Staatsdruckerei, 1911): xlvi–lxi.

K.k. Statistische Zentralkommission, ed. "Gesamtverkehr auf den österreichischen Flüssen in den Jahren 1908, 1909, und 1910." *Oesterreichische Statistik* 93, no. 3 (K.k. Hof- und Staatsdruckerei, 1916): x–xi.

K.k. Statistische Zentralkommission, ed. *Österreichisches Statistisches Handbuch: Sechsundzwanzigster Jahrgang 1907*. Verlag der k.k. statistischen Zentralkommission, 1908.

Klein, Christine A., and Sandra B. Zellmer. *Mississippi River Tragedies: A Century of Unnatural Disaster*. New York University Press, 2014.

Kleinschrod, Carl Theodor von. *Die Kanal-Verbindung des Rheins mit der Donau*. Franz, 1834.

Klun, Vinzenz. *Allgemeine Geographie mit besonderer Rücksicht auf das Kaiserthum Oesterreich*. C. Gerold's Sohn, 1861.

Klun, Vinzenz. "Flusskarten der Donau und der Theiss." In *Mittheilung der kaiserlich-königlichen Geographischen Gesellschaft VII. Jahrgang 1863*, edited by Franz Foetterle, 1–17. F. B. Geitler, 1863.

"Klun, Vinzenz." In *Österreichisches Biographisches Lexikon 1815–1950*. 16 vols. Verlag der Österreichischen Akademie der Wissenschaften, 1965.

Kohl, Johann Georg. *Die Donau von ihrem Ursprunge bis Pesth*. Trieste: Verlag der literarisch-artistischen Abtheilung der Oesterreichischen Lloyd, 1854.

Komlos, John. *The Habsburg Monarchy as a Customs Union: Economic Development in Austria-Hungary in the Nineteenth Century*. Princeton University Press, 1983.

Komlosy, Andrea. "Innere Peripherien im räumlichen Mehrebenensystem: Das habsburgische Beispiel im 19. und frühen 20. Jahrhundert." *Österreichische Zeitschrift für Geschichtswissenschaften* 31, no. 2 (2020): 95–124.

Selected Bibliography

Koselleck, Reinhart. "Einleitung." In *Geschichtliche Grundbegriffe. Historisches Lexikon zur politisch-sozialen Sprache in Deutschland*, edited by Otto Brunner, Werner Conze, and Reinhart Koselleck, xiii–xxvii. Ernst Klett, 1972.

Kożuchowski, Adam. *The Afterlife of Austria-Hungary: The Image of the Habsburg Monarchy in Interwar Europe*. University of Pittsburgh Press, 2013.

Krejci, Heinz. *Expedition in die Kulturegeschichte des Abwassers*. Bösmüller, 2004.

Kreuter, Franz. *Praktisches Handbuch der Drainage*. Carl Gerold und Sohn, 1854.

Lackner, Helmut, and Gerhard A. Stadler. *Fabriken in der Stadt: Eine Industriegeschichte der Stadt Linz*. Archiv der Stadt Linz, 1990.

Landesgesetz- und Regierungsblatt für das Erzherzogthum Oesterreich unter der Enns. K.k. Hof- und Staatsdruckerei, 1852.

Landgraf, János. "Halászat." In *Magyarország földművelése 1896*, edited by Földművelésügyi M. Kir. Ministerium, 391–408. Viktor Hornyánszky, 1896.

Landry, Marc, and Patrick Kupper, eds. *Austrian Environmental History*. University of New Orleans Press, 2018.

Lászlóffy-Böhm, Woldemár. "A Tiszavölgy: Vízrajzi leírás és a vízimunkálatok ismertetése 1833–1932." *Vízügyi Közlemények* 14, no. 2 (1932): 108–42.

Laurencic, Julius, ed. "In Linz und Gastein." *Österreich in Wort und Bild: Vaterländisches Jubiläums-Prachtwerk*. Vienna: Georg Szelinski, o.J., 1898.

Lenner, Emil. *A győri m. kir. állami főreáliskola huszonkettedik évi értesítője az 1894–95 tanévről*. Gross Testvérek Könyvnyomtató Intézetéből, 1895.

Linde, Franz Xaver. *Chronik des Marktes und der Stadt Melk umfassend den Zeitraum von 890 bis 1899 mit besonderer Berücksichtigung der letzten 34 Jahre*. Selbstverlag der Gemeinde Melk, 1900.

Lindström, Fredrik. *Empire and Identity: Biographies of the Austrian State Problem in the Late Habsburg Empire*. Purdue University Press, 2008.

Lóczy, Dénes. "The Danube: Morphology, Evolution and Environmental Issues." In *Large River Geomorphology and Management*, edited by Avijit Gupta, 235–60. John Wiley, 2007.

Loessi, Friedrich Ritter von. *Technischer Bericht zum Projecte eines Schiffahrts-Kanales zwischen der Save und Donau in der Militärgrenze und Slavonien*. Kränzel, 1869.

"Lower Danube Green Corridor: Floodplain Restoration for Flood Protection." *Climate-ADAPT*, June 7, 2016. https://climate-adapt.eea.europa.eu/en/metadata/case-studies/lower-danube-green-corridor-floodplain-restoration-for-flood-protection.

Selected Bibliography

Luckin, Bill, Dieter Schott, and Geneviére Massard-Guilbaud, eds. *Resources of the City: Contributions to an Environmental History of Modern Europe*. Routledge, 2005.

Lukács, Béla. "Közlekedési intézmények." In Vol. 5/2, *Az Osztrák-Magyar Monarchia írásban és képekben. Magyarország I/2*. A Magvar Királyi Államnyomda Kiadás, 1888.

Maclean, Donald. *Remarks on the Facility Which Steam Navigation Affords for Invading Great Britain and Ireland*. James Ridgway, 1824.

Magyar Statistikai Évkönyv: tizennyolcadik évfolyam, 1888. Az Athenaeum Irodalmi és Nyomdai R. Társulat Könyvnyomda, 1891.

Major, Jószef. "A Ráckevei (Soroksári)- Duna szabályozásának fejlődése és helyzete." In *Árvízvédelem, Folyó- és Tószabályozás, Víziutak Magyarországon*, edited by Dezső Kovács. Országos Vizügyi Hivatal, 1978.

Malm, Andrea. *Fossil Capital: The Rise of Steam Power and the Roots of Global Warming*. Verso, 2016.

Malte-Brun, Conrad. *Universal Geography, Or, A Description of All Parts of the World, on a New Plan According to the Great Natural Divisions of the Globe: Accompanied with Analytical, Synoptical, and Elementary Tables: Improved by the Addition of the Most Recent Information*. John Laval and S. F. Bradford, 1829.

Mann, Theodore Augustin. *A Treatise on Rivers and Canals*. J. Nichols, 1780.

Matis, Herbert, and Karl Bachinger. "Österreichs industrielle Entwicklung." In *Die Hasburgmonarchie, 1848–1914 Band I: Wirtschaftliche Entwicklung*, edited by Alois Brusatti, 105–232. Austrian Academy of Sciences, 1973.

Mauch, Christoph. "Introduction." In Mauch and Pfister, *Natural Disasters, Cultural Responses*, 1–16.

Mauch, Christof, and Christian Pfister, eds. *Natural Disasters, Cultural Responses: Case Studies Toward a Global Environmental History*. Lexington Books, 2009.

Mauch, Christof, and Thomas Zeller. "Rivers in History and Historiography: An Introduction." In Mauch and Zeller, *Rivers in History*, 1–10.

Mauch, Christof, and Thomas Zeller, eds. *Rivers in History: Perspectives on Waterways in Europe and North America*. University of Pittsburgh Press, 2011.

Mauelshagen, Franz. "Disaster and political culture in Germany since 1500." In Mauch and Pfister, *Natural Disasters, Cultural Responses*, 41–75.

Mayrhofer, Fritz, and Willibad Katzinger. *Geschichte der Stadt Linz: Band II: Von der Aufklärung zur Gegenwart*. Verlag J. Wimner, 1990.

McCarney-Castle, Kerry, George Voulgaris, Albert J. Kettner, and Liviu Giosan. "Simulating Fluvial Fluxes in the Danube Watershed: The 'Little Ice Age' Versus Modern Day." *Holocene* 22, no. 1 (2011): 91–105.

Selected Bibliography

McNeill, J. R. "Observations on the Nature and Culture of Environmental History." *History and Theory* 42, no. 4 (2003): 5–43.

McNeill, J. R. *Something New Under the Sun: An Environmental History of the Twentieth-Century World*. W. W. Norton, 2000.

McNeill, J. R., and Peter Engelke. *The Great Acceleration: An Environmental History of the Anthropocene Since 1945*. Harvard University Press, 2014.

McNeill, J. R., and William McNeill. *The Human Web: A Bird's-Eye View of World History*. W. W. Norton, 2003.

Meissinger, Otto. *Historische Donauschiffahrt: Holzschiffe u. Flösse*. Verlag Kurt Wedl, 1975.

Melo, Marián, Pavla Pekárová, Pavol Miklánek, Katarína Melová, and Cyntia Dujsíková. "Use of Historical Sources in a Study of the 1895 Floods on the Danube River and Its Tributaries." *Geographica Pannonica* 18, no 4 (2014): 108–16.

Mészáros, Vince. *Gróf Széchenyi István al-dunai diplomáciai kapcsolatai* [Count Széchenyi's Lower Danube Diplomatic Relations]. Magyar Vízügyi Múzeum, 1991.

Mevissen, Robert Shields. "Forged in the Floods: Transnational Networks in the Habsburg Monarchy." *Water History* 12 (2020): 265–80.

Mevissen, Robert Shields. "Meandering Circumstances, Fluid Associations: Shaping Riverine Transformations in the Late Habsburg Monarchy." *Austrian History Yearbook* 49 (2018): 23–40.

Miller, Paul, and Claire Morelon, eds. *Embers of Empire: Continuity and Rupture in the Habsburg Successor States After 1918*. Berghahn Books, 2019.

Mitchell, A. Wess. *The Grand Strategy of the Habsburg Empire*. Princeton University Press, 2018.

Mitchell, B. R., ed. *European Historical Statistics, 1750–1970*. Palgrave Macmillan, 1975.

Mittas, Sofie. "Reconstruction and Transformation of the Austrian Wood-Paper Commodity Chain, 1945–1955." In Landry and Kupper, *Austrian Environmental History*, 133–55.

Mokre, Jan. "The Environs Map: Vienna and Its Surroundings c. 1600–c. 1850." *Imago Mundi* 49 (1997): 90–103.

Müller, Adelbert. *Die Donau vom Ursprunge bis zu den Mündungen. Zugleich ein Handbuch für Reisende, welche diesen Strom befahren*. Georg Joseph Manz, 1839.

Murray, John III. *A Handbook for Travellers in Southern Germany*. 2nd ed. Johan Murray and Son, 1840.

Mutschlechner, Martin. "Die Gleichzeitigkeit des Ungleichzeitigen: Alphabetisierung als Gradmesser der Entwicklung." *Die Welt der Habsburger*. https://ww1.habsburger.net/de/kapitel/die-gleichzeitigkeit-des-ungleichzeitigen-alphabetisierung-als-gradmesser-der-entwicklung.

Selected Bibliography

Nagy, László. *Az 1876 évi árvízek*. Környezetvédelmi és Vízügyi Minisztérium, 2007.

Naimark, Norman. *Fires of Hatred: Ethnic Cleansing in Twentieth-Century Europe*. Harvard University Press, 2002.

Nemes, Robert. *Another Hungary: The Nineteenth-Century Provinces in Eight Lives*. Stanford University Press, 2016.

Nemes, Robert. *The Once and Future Budapest*. Northern Illinois University Press, 2005.

Neundlinger, Michael. "Disaster Ahead: How Danube Floods Created Telegraph Networks." *Arcadia in Development: Online Explorations of Global Environmental History* (2012). http://www.environmentandsociety.org/arcadia/disaster-ahead-how-danube-floods-created-telegraph-networks.

Neweklowsky, Ernst. "Die Donau bei Linz und ihre Regelung." *Naturkundliches Jahrbuch d. Stadt Linz 1955* (Linz, 1955): 171–226.

Oberösterreichischer Landesausschuss, ed. *Bericht über die Thätigkeit des oberösterreichischen Landtages und des von diesem gewählten Landesausschusses in der VII. Wahlperiode vom 15. September 1884 bis Sommer 1890*. Verlag des Landesausschuss, 1890.

Oberösterreichischer Landesausschuss, ed. *Bericht über die Thätigkeit des oberösterreichischen Landtages und des von diesem gewählten Landesausschusses in der VIII. Wahlperiode vom 14. Oktober 1890 bis Sommer 1896*. Verlag des Landesausschuss, 1896.

Oberösterreichischer Landesausschuss, ed. *Bericht über die Thätigkeit des oberösterreichischen Landtages und des von diesem gewählten Landesausschusses in der IX. Wahlperiode vom Sommer 1896 bis Sommer 1902*. Verlag des Landesausschuss, 1902.

Oberösterreichischer Landesausschuss, ed. *Bericht über die Thätigkeit des oberösterreichischen Landtages und des von diesem gewählten Landesausschusses in der X. Wahlperiode vom Sommer 1902 bis Herbst 1908*. Verlag des Landesausschuss, 1908.

Oelwein, Arthur. *Ausbau der Wasserstrassen in Mittel-Europa: Zwei Vorträge, gehalten am 6. December 1881 und am 31. Januar 1882 im Club österreichischer Eisenbahn-Beamten von Arthur Oelwein, Bau-Inspector der k.k. Direction für Staatseisenbahn-Betrieb in Wien*. Lehmann & Wentzel, 1882.

Oelwein, Arthur. *Die Binnen-Wasserstrassen im Transportgeschäfte der Gegenwart. Vortag gehalten im Niederosterr: Gewerbeverein am 6. November 1891 von k.k. Professor Arthur Oelwein*. Verlag des Niederoesterreichischen Gewerbevereins, 1891.

Oelwein, Arthur. *Die Regulirung der Donau nächst Linz und die Anlage eines Hafens daselbst*. Linz: Joseph Wimmer, 1882.

Oelwein, Arthur. *Die Wasserstraßenfrage in Oesterreich*. Gerold, 1894.

Okey, Robin. *The Habsburg Monarchy, c. 1765–1918*. Palgrave, 2001.

Selected Bibliography

Osterhammel, Jürgen. *The Transformation of the World: A Global History of the Nineteenth Century.* Translated by Patrick Camiller. Princeton University Press, 2014.

Österr.-ungar. *Blätter für Geflügel- und Kaninchenzucht dann für Bienen- und Fischzucht, Sing- und Ziervögel-Pflege.* J. H. Nowotny, 1879.

Otruba, Gustav. "Linz, seine neue Strafanstalt, die Messingfabrik im Schloß Lichtenegg bei Wels und die Wollenzeugfabrik in Linz." *Oberösterreichische Heimatsblätter,* vol. 4, 295–318. Landesinstitut für Volksbildung und Heimatpflege in Oberösterreich, 1989.

Pach, Zsigmond Pál. "Széchenyi és az Alduna-szabályozás 1830–1832-ben." *Történelmi szemle* 18, no. 4 (1975): 557–82.

Paget, John. *Hungary and Transylvania; with Remarks on Their Condition, Social, Political, and Economical.* Vol. 1. John Murray, 1850.

Palacký, František. "Letter to Frankfurt, 11 April 1848." In *National Romanticism: The Formation of National Movements Discourses of Collective Identity in Central and Southeast Europe 1770–1945,* vol. 2, edited by Balázs Trencsényi and Michal Kopeček, 322–29. Central European University Press, 2013.

Parker, Geoffrey. *Global Crisis: War, Climate Change and Catastrophe in the Seventeenth Century.* Yale University Press, 2013.

Pasetti, Ritter von. *Notizen über die Donauregulierung im österreichischen Kaiserstaate bis zum Ende des Jahre 1861 mit Bezug auf die im k.k. Staatsministerium herausgegebenen Übersichtskarte der Donau.* k.k. Hof- und Staattsdruckerei, 1862.

Paulmann, Johannes. "The Straits of Europe: History at the Margins of a Continent." *Bulletin of the German Historical Institute* 52 (2013): 7–28.

Peez, Dr. Alexander. *II. internationaler Binnenschiffahrts-Congress Wien 1886: I. Section: Der wirthschaftliche Werth der Binnen-Wasserstrassen.* Verlag der Organisations-Commission des Congresses, 1886.

Pfister, Christian, Rudolf Brázdil, Jürg Luterbacher, Astrid E. J. Ogilvie, and Sam White. "Early Modern Europe." In White, Pfister, and Mauelshagen, *Palgrave Handbook of Climate History,* 265–95.

Pilsitz, Martin. "Die Donau als Faktor der industriellen Stadtentwicklung in Pest im 19. Jahrhundert." In Tamáska and Szabó, *Donau-Stadt-Landschaften,* 195–209.

Pinke, Zsolt. "Modernization and Decline: An Eco-Historical Perspective on Regulation of the Tisza Valley, Hungary." *Journal of Historical Geography* 45 (2014): 92–105.

Pisecky, Franz. "Die Grösste Binnenreederei der Welt: 140 Jahre Erste Donau-Dampfschiffahrts-Gesellschaft- Größe und europäische Bedeutung der österreichischen Donauschiffahrt." *Tradition: Zeitschrift für Firmengeschichte und Unternehmerbiographie,* vol. 2/3 (1970): 49–66.

Selected Bibliography

Planche, J. R. *Descent of the Danube from Ratisbon to Vienna, During the Autumn of 1827 with Anecdotes and Recollections, Historical and Legendary, of the Towns, Castles, Monasteries, &c, Upon the Banks of the River, and their Inhabitants and Proprietors, Ancient and Modern.* James Duncan, 1828.

Pomeranz, Kenneth. *The Great Divergence: China, Europe, and the Making of the Modern World Economy.* Princeton University Press, 2000.

Popper, Otto. "The International Regime of the Danube." *Geographical Journal* 102, no. 5/6 (1943): 240–53.

Prechtl, Johann Joseph. "Zur Geschichte der Dampfboote." In *Jahrbücher des kaiserlichen-königlichen polytechnischen Institutes in Wien*, edited by Johann Joseph Prechtl, 208–17. Vol. 1. Carl Gerold, 1819.

Pritchard, Sara B. *Confluence: The Nature of Technology and the Remaking of the Rhône.* Harvard University Press, 2011.

Prónay, Gábor. *Skizzen aus dem Volksleben in Ungarn.* Hermann Geibel, 1855.

Quinn, Michael J. *A Steam Voyage down the Danube: With Sketches of Hungary, Wallachia, Servia, Turkey, Etc.* 3rd London ed. Theodore Foster, 1836.

Rácz, Lajos. "The Climate History of Central Europe in the Modern Age." In *People and Nature in Historical Perspective*, edited by József Laszlovszky and Péter Szabó, 229–46. PXP Rt, 2003.

Rácz, Lajos. "The Danube Pontoon Bridge of Pest-Buda (1767–1849) as an Indicator and Victim of the Climate Change of the Little Ice Age." *Global Environment* 9, no. 2 (2016): 458–83.

Rácz, Lajos. *The Steppe to Europe: An Environmental History of Hungary in the Traditional Age.* Translated by Alan Campbell. White Horse Press, 2013.

Rechenschaftsbericht des Gemeinderathes der Landeshauptstadt Linz über seine Thätigkeit im Jahre 1899 nebst anderen statistischen Daten. Selbstverlag des Gemeinderathes, 1900.

Rechenschaftsbericht des Gemeinderathes der Landeshauptstadt Linz über seine Thätigkeit im Jahre 1907 nebst anderen statistischen Daten. Selbstverlag des Gemeinderathes, 1908.

Rechenschaftsbericht des Gemeinderathes der Landeshauptstadt Linz über seine Thätigkeit im Jahre 1913 nebst anderen statistischen Daten. Selbstverlag des Gemeinderathes, 1914.

Reise auf der Donau von Ulm bis Wien, mit Angabe aller Städte, Flecken, Dörfer, Schlösser, u. an beyden Ufern, ihrer vornehmsten Merkwürdigkeiten, und der Flüsse, welche sich mit der Donau vereinigen. Beck'schen Buchhandlung, 1813.

Reitter, Ferenc. *Donau-Regulierung zwischen Pest und Ofen, Pester Schiffahrts-Canal, Schutz der Insel Csepel und des linkseitigen Ufers des Soroksárer Donauarmes gegen Ueberschwemmung: Drei Anträge.* Gebrüder Pollack, 1865.

Renan, Ernest. "What Is a Nation?" Sorbonne conference, March 11, 1882. Paris.

Selected Bibliography

Rhodes, Richard. *Energy: A Human History.* Simon and Schuster Paperbacks, 2018.

Ricci, Matteo Ricci. *China in the Sixteenth Century: The Journals of Matthew Ricci, 1583–1610.* Translated by Louis J. Gallagher. Random House, 1942.

Rohr, Christian. "The Danube Floods and Their Human Response and Perception (14th to 17th C)." *History of Meteorology* 2 (2005): 71–86.

Romsics, Ignác. "The Peasantry and the Age of Revolutions: Hungary, 1918–1919." *Acta Historica Academiae Scientiarum Hungaricae* 35 (1989): 113–33.

Ross, Corey. *Power and Ecology in the Age of Empire: Europe and the Transformation of the Tropical World.* Oxford University Press, 2017.

Rozsnyai, József. "Industrial Buildings and Halls." In *Motherland and Progress: Hungarian Architecture and Design, 1800–1900*, edited by József Sisa, 637–51. Birkhäuser, 2016.

Ruprecht, M. "Bericht über die Entwässerungsarbeiten in der Insel Schütt." In *Verhandlung des Vereins für Naturkunde zu Presburg*, edited by E. Mack, 34–41. C. F. Wigand, 1865.

Ruß, Dr. Viktor. *Der volkswirtschaftliche Wert der künstlichen Schiffahrtsstraßen.* Verlagsbuchhandlung Georg D. W. Callwey, 1901.

Sárközi, Zoltán. *Árvizek, ármentesítés és folyószabályozás a szigetközben és az alsó-Rába vidékén.* Budapesti Műszaki Egyetem Központi Könyvtára Műszaki Tudománytörténeti Kiadványok, 1968.

Sartori, Franz. *Wien's Tage der Gefahr und die Retter aus der Noth, eine authentische Beschreibung der unerhörten Ueberschwemmung des flachen an der Donau gelegenen Landes in Oesterreich unter der Enns.* Carl Gerold, 1832.

Schediwy, Robert. *Städtebilder: Reflexionen zum Wandel in Architektur und Urbanistik.* LIT Verlag, 2005.

"Schiffverkehr auf der oberen Donau." *Statistische Monatsschrift* 1 (Alfred Hölder, 1875): 34.

Schmid, Martin, Gertrud Haidvogl, Thomas Friedrich, Andrea Funk, Lisa Schmalfuss, Astrid Schmidt-Kloiber, and Thomas Hein. "The Danube: On the Environmental History, Present, and Future of a Great European River." In *River Culture: Life as a Dance to the Rhythm of the Waters*, edited by Karl M. Wantzen, 637–71. UNESCO Publishing, 2023.

Schott, Dieter. "Urban Environmental History: What Lessons Are There to Be Learned?" *Boreal Environmental History* 5 (2004): 519–28.

Schriften der in Budapest am 4. September des Jahres 1916 abgehaltenen Donaukonferenz. "Pátria" literar. Unternehmen und Druckerei Aktien-Gesellschaft, 1916.

Schultes, Dr. J. A. *Donau-Fahrten: Ein Handbuch für Reisende auf der Donau.* Vol. 2. J. G. Cotta'schen Buchhandlung, 1827.

Selected Bibliography

Schweiger-Lerchenfeld, Amand von. *Die Donau als Völkerweg, Schiffahrtstrasse, und Reiseroute*. R. Hartleben's Verlag, 1896.

Scott, James C. *Seeing Like a State: How Certain Schemes to Improve the Human Condition Have Failed*. Yale University Press, 1998.

Šedivý, Miroslav. "Metternich and the Suez Canal: Informal Diplomacy in the Interests of Central Europe." *Central European History* 55 (2022): 372–89.

Shedel, James. "The Elusive Fatherland: Dynasty, State, Identity and the Kronprinzenwerk." In *Inszenierung des kollektiven Gedächtnisses: Eigenbilder, Fremdbilder* edited by Moritz Csáky and Klaus Zeyringer, 70–82. StudienVerlag, 2002.

Shedel, James. "Emperor, Church, and People: Religion and Dynastic Loyalty During the Golden Jubilee of Franz Joseph." *Catholic Historical Review* 76 (1990): 71–92.

Shedel, James. "*Fin de Siècle* or *Jahrhundertwende*: The Question of an Austrian Sonderweg." In *Rethinking Vienna: 1900*, edited by Steven Beller, 80–104. Berghahn Books, 2001.

Shedel, James. "The Mother of It All: Maria Theresia and the Creation of the Hybrid Monarchy." In *Marija Terezija: Med razsvetljenskimi reformami in zgodovinskim spominom*, edited by Miha Preinfalk and Boris Golec, 9–18. Založba ZRC, 2019.

"Ship Canals in Austria." *Geographical Journal* 18, no. 3 (1901): 289–91.

Simonkay, Márton. "Felső-Duna-völgy dualizmuskori szabályozása." Bachelor's thesis, Eötvös Loránd University, 2013.

Simonkay, Marton. "'*Vagy ilyen szabályozás lesz, vagy semmilyen*' A Rábaszabályozó Társulat első évtizede (1873–1883)." Unpublished manuscript, 2013.

Sisa, József. "Count Ferenc Széchényi's Visit to English Parks and Gardens in 1787." *Garden History* 22, no. 1 (1994): 64–71.

Sitzungs-Protokoll der General-Versammlung der k.k. priv. ersten Donau-Dampfschiffahrts-Gesellschaft vom 10. Februar 1840. Selbstverlag, 1840.

Sitzungs-Protokoll der General-Versammlung der k.k. priv. ersten Donau-Dampfschiffahrts-Gesellschaft vom 29. Jänner 1838. A. Strauß's fel. Witwe, 1838.

Sked, Alan. *The Decline and Fall of the Habsburg Monarchy, 1815–1918*. 2nd ed. Longman, 2001.

Smil, Vaclav. *Energy and Civilization: A History*. MIT Press, 2017.

Smil, Vaclav. *Energy Transitions: Global and National Perspectives*. 2nd ed. Praeger, 2017.

Smith, Theodore. "Water and Death." *Sanitary Era: Progressive Health Journal* 3, no. 12 (1889): 172.

Soden, Friedrich Julius Heinrich Graf von. *Der Maximilians-Canal: Über die*

Selected Bibliography

Vereinigung der Donau mit dem Main und Rhein. Mit 1 Karte. Riegel und Wießner, 1822.

Solymos, Ede. *Élet a Dunán: A halászok, vízen járók élete Baján és környékén.* Bajai Türr István Múzeum, 2004.

Sommer, Johann Gottfried. *Das Kaiserthum Oesterreich, geographisch-statistisch dargestellt.* J. G. Calve'sche Buchhandlung, 1839.

Statistischer Bericht über die gesammten wirtschaftlichen Verhältnisse Oberösterreichs in den Jahren 1876–1880. Verlag der oberösterreichischen Handels- und Gewerbekammer, 1881.

Steidl, Annemarie, Engelbert Stockhammer, and Hermann Zeitlhofer. "Relations Among Internal, Continental, and Transatlantic Migration in Late Imperial Austria." *Social Science History* 31, no. 1 (2007): 61–92.

Stenographische Protokolle über die Sitzungen des Hauses der Abgeordneten des österreichischen Reichsrathes im Jahre 1901. XVII Session, vol. 3. K.k. Hof- und Staatsdruckerei, 1901.

Stenographische Protokolle über die Sitzungen des Hauses der Abgeordneten des österreichischen Reichsrathes in den Jahren 1886 und 1887. X. Session, vol. 4. K.k. Hof- und Staatsdruckerei, 1887.

Stenographische Protokolle über die Sitzungen des Hauses der Abgeordneten des österreichischen Reichsrathes. VII. Session, vol. 2. K.k. Hof- und Staatsdruckerei, 1873.

Stenographisches Protokoll der Versammlung des Donau-Vereines zur gemeinschaftlichen Berathung mit dem Gemeinderathe der Stadt Linz am 15. Juni 1884. Linz: Verlag des Gemeinderathes der Landeshauptstadt Linz, 1884.

Stephanov, Darin. "Ruler Visibility, Modernity, and Ethnonationalism in the Late Ottoman Empire." In *Living in the Ottoman Realm: Empire and Identity, 13th to 20th Centuries,* edited by Christine Isom-Verhaaren and Kent F. Schull, 259–71. Indiana University Press, 2016.

Stix, Edmund, ed. *Zeitschrift des österreichischen Inginieur- und Architekten-Vereins.* Druck und Verlag der artischen Anstalt von R. v. Waldheim, 1871.

Strickland, Lloyd, and Michael Church. "Leibnitz's Observations on Hydrology: An Unpublished Letter on the Great Lombardy Flood of 1705." *Annals of Science* 72, no. 4 (2015): 517–32.

Suess, Eduard. *Die Aufgabe der Donau.* Druck von A. Schaft im Verlag des Reform-Vereins der Wiener Kaufleute, 1880.

Suess, Eduard. *Errinerungen.* S. Hirzel, 1916.

Sugar, Peter F., Péter Hanák, and Tibor Frank, eds. *A History of Hungary.* Indiana University Press, 1994.

Szabadfalvi, József. "Nomadic Wintering System on the Great Hungarian Plain." *Acta Ethnographica Academiae Scientiarum Hungaricae Tomus* 17 (1968): 139–67.

Selected Bibliography

Szabó, Csaba. "Brücken über die Donau zwischen Ofen und Pest: Kettenbrücke, Margaretenbrücke, Franz-Joseph-Brücke, Elisabethbrücke." In Csendes and Sipos, *Budapest und Wien*, 89–105.

Szabó, Julianna. "Budapester Stadtpanorama mit der Donau: Zeitgenössische Fragen in historischem Kontext." In Tamáska and Szabó, *Donau-Stadt-Landschaften*, 135–51.

Szász, József. *Die ungarische Landwirtschaft der Gegenwart mit besonderer Berücksichtigung der Extensität und Intensität ihres Betriebes*. Friedrich-Wilhelms-Universität Buchdruckerei, 1907.

Széchenyi, István. *Über die Donauschifffahrt*. Translated by Michael v. Paziazi. Johann Gyurián and Martin Bagó, 1836.

Tafeln zur Statistik der österreichischen Monarchie, 3. Jahrgang 1831.

Tafeln zur Statistik der österreichischen Monarchie, 10. Jahrgang 1837.

Tafeln zur Statistik der österreichischen Monarchie, 11. Jahrgang 1838.

Tafeln zur Statistik der österreichischen Monarchie, 12. Jahrgang 1839.

Tafeln zur Statistik der österreichischen Monarchie, 13. Jahrgang 1840.

Tamáska, Máté, and Csaba Szabó, eds. *Donau-Stadt-Landschaften, Danube-City-Landscapes*. Lit Verlag, 2016.

Tarr, Joel A. *The Search for the Ultimate Sink: Urban Pollution in Historical Perspective*. University of Akron Press, 1996.

Timms, Edward. "National Memory and the 'Austrian Idea' from Metterich to Waldheim." *Modern Language Review* 86, no. 4 (1991): 898–910.

Tirscher, Pál. "A hajózás és forgalom a Dunán és mellékfolyóin 1865/8-ban." In *Statisztikai és Nemzetgazdasági Közlemények: A hazai ismeretének előmozdítására* 5/1, edited by János Hunfalvy, 195–231. Magyar Akad. Könyvtár, 1868.

Tóth, Csaba, Tibor Novák, and János Rakonczai. "Hortobágy Puszta: Microtopography of Alkali Flats." In *Landscapes and Landforms of Hungary*, edited by Dénes Lóczy, 237–46. Springer, 2015).

Trautsamwieser, Herbert. *Weisse Schiffe am Blauen Strom*. Malek Verlag, 1996.

Tredgold, Thomas. *Remarks on Steam Navigation, and Its Protection, Regulation and Encouragement. In a Letter to the Right Honourable W. Huskisson, etc.* Longman, Hurst, Reese, Orme, Brown, and Green, 1825.

Truesdell, Matthew. *Spectacular Politics: Louis-Napoleon Bonaparte and the Fête Impériale, 1849–1870*. Oxford University Press, 1997.

Tschischka, Franz. *Der Gefährte auf Reisen in dem österreichischen Kaiserstaate*. Friedrich Beck's Universitäts-Buchhandlung, 1834.

Uekoetter, Frank, ed. *The Turning Points of Environmental History*. University of Pittsburgh Press, 2010.

Unowsky, Daniel. *The Pomp and Politics of Patriotism: Imperial Celebrations in Habsburg Austria, 1848–1916*. Purdue University Press, 2005.

Vadas, András, and László Ferenczi. "Small Urban Waters and Environmental

Selected Bibliography

Pressure Before Industrialization: The Case of Hungary." *Journal of Historical Geography* 82 (2023): 98–109.

Vadas, Ferenc. "Die Regulierung der Donau und die Kaianlagen." In Csendes and Sipos, *Budapest und Wien*, 79–87.

Vari, Alexander. "From 'Paris of the East' to 'Queen of the Danube': International Models in the Promotion of Budapest Tourism, 1885–1940." In *Touring Beyond the Nation: A Transnational Approach to European Tourism History*, edited by Eric G. E. Zuelow, 103–26. Ashgate, 2011.

Vári, András. "The Functions of Ethnic Stereotypes in Austria and Hungary in the Early Nineteenth Century." In *Creating the Other: Ethnic Conflict and Nationalism in Habsburg Central Europe*, edited by Nancy M. Wingfield, 39–55. Berghahn Books, 2003.

Vedres, István. *Ueber einen neuen Schiffbaren Kanal in Ungerland, mitelst dessen die Donau mit der Theiß am vortheilhaftesten verbunden werden kann*. Translated by Nikolaus Stankovitsch. Grünn, 1805.

Veichtlbauer, Ortrun. "Port of Vienna: Infrastructures and War on the Danube River in Vienna, 1850–1950." In Landry and Kupper, *Austrian Environmental History*, 73–101.

Veichtlbauer, Ortrun. "Von der Strombaukunst zur Stauseenkette: Die Regulierung der Donau." In Winiwarter and Schmid, *Umwelt Donau*, 57–74.

Veres, Madalina Valeria. "Constructing Imperial Spaces: Habsburg Cartography in the Age of Enlightenment." PhD diss., University of Pittsburgh, 2015.

Verordnungs-Blatt für den Dienstbereich österreichischen Finanzministeriums, Jahrgang 1854. K.k. Hof- und Staatsdruckerei, 1854.

Verordnungs-Blatt für den Dienstbereich österreichischen Finanzministeriums, Jahrgang 1858. K.k. Hof- und Staatsdruckerei, 1858.

Viner, Jacob. "Adam Smith and Laissez Faire." *Journal of Political Economy* 35, no. 2 (1927): 198–232.

Voltaire. *The History of Peter the Great, Emperor of Russia*. Translated by Tobias Smollett. S. Johnson & Son, 1845.

Wagar, Chip. *Double Emperor: The Life and Times of Francis of Austria*. Hamilton Books, 2018.

Walcher, Joseph. *Nachrichten von den bis auf das Jahr 1791 an dem Donau-Strudel zur Sicherheit der Schiffahrt fortgesetzen Arbeiten, nebst einem Anhange von der physikalischen Beschaffenheit des Donau-Wirbels*. Joseph Edlen von Kurzbeck, 1791.

Wallace, Sigismund. *Auf der Donau von Wien nach Constantinopel und nach den Dardanellen*. I. C. Zamarski & C. Dittmarsch, 1864.

"Was ist die Hydrographie Österreichs?" *Ministerium für ein Lebenwertes Österreich*. https://info.bml.gv.at/themen/wasser/wasser-oesterreich/hydrographie/Organisation_HZB.html.

Selected Bibliography

Watznetter, Walter. "Danube Islanders: Population Growth and Social Change in Vienna's Second and Twentieth Districts, from the Regulation of the Danube to Current Patterns of Gentrification." In Tamáska and Szabó, *Donau-Stadt-Landschaften*, 87–98.

Wex, Gustav Ritter von. *Lecture on the Improvement of the Danube at Vienna, Delivered Before the Society of Austrian Engineers and Architects on March 18, 1876*. Translated by G. Weitzel. Government Printing Office, 1880.

White, Richard. *The Organic Machine: The Remaking of the Columbia River*. Hill and Wang, 1995.

White, Sam, Christian Pfister, and Franz Mauelshagen, eds. *The Palgrave Handbook of Climate History*. Palgrave, 2018.

Wilson, Samuel J. "Lost Opportunities: Lajos Kossuth, the Balkan Nationalities, and the Danubian Confederation." *Hungarian Studies* 8/2 (1993): 171–93.

Winckler, J. "Der Wiener Donauhandel bis zum Jahre 1874." *Statistische Monatsschrift* 2 (1876): 1–23.

Winckler, Johann. *Übersicht des Schiffs- und Waarenverkehrs auf der oberen Donau zu Wien, Linz, und Engelhartszell in den Jahren 1849–1869*. k.u.k. Hof- und Staatsdruckerei, 1870.

Winiwarter, Verena, and Martin Schmid, eds. *Umwelt Donau: Eine andere Geschichte*. Niederösterreichisches Landesarchiv, 2010.

Wood, Nathaniel D. "Not Just the National: Modernity and the Myth of Europe in the Capital Cities of Central and Southeastern Europe." In *Capital Cities in the Aftermath of Empires: Planning in Central and Southeastern Europe*, edited by Emily Gunzburger Makaš and Tanja Damljanović Conley, 258–69. Routledge, 2010.

Worster, Donald. *Rivers of Empire: Water, Aridity, and the Growth of the American West*. Pantheon Books, 1985.

Wunderl, Stefan. "Die Geschichte der Schiffswerft Korneuburg unter Berücksichtigung der Situation der Arbeiterschaft." MA thesis, University of Vienna, 2008.

Zahra, Tara. *Kidnapped Souls: National Indifference and the Battle for Children in the Bohemian Lands, 1900–1948*. Cornell University Press, 2012.

Zeilinger, Elisabeth. "Zur Urheberschaft der Pasetti-Karte: Die Rolle von Valentin von Streffleur als Initiator der berühmten Donau-Karte." Research Blog. Österreichische Nationalbibliothek, August 27, 2021. https://www.onb.ac.at/forschung/forschungsblog/artikel/zur-urheberschaft-der-pasetti-karte.

Zels, Louis. *Die Activen der Ersten k.k. priv. Donau-Dampfschifffahrts-Gesellschaft*. Spielhagen & Schurich, 1891.

Zels, Louis. *Die Regulirungskosten der Donau*. Fromme, 1880.

Zels, Louis. *Die Selbstkosten des Eisenbahn-Transportes und die Masserstrassen-*

Selected Bibliography

Frage: Eine Polemik gegen das gleichnamige Buch des Wilhelm Ritter von Nördling. Spielhagen u. Schurich,1886.

Zels, Louis. *Schiffahrtscanal-Projekte aus der Josephinischen-Zeit und deren Verwendbarkeit für die Gegenwart: Vortrag, gehalten im Wiener kaufmännischen Vereine am 20.December 1880.* Selbstverlag, 1882.

Zels, Louis. *Über Wasserstrassen: Vortrag.* Spielhagen u. Schurich, 1887.

Zels, Louis. *Versuch einer Statistik des Betriebes der Ersten k.k. priv. Donau-Dampfschiffahrts-Gesellschaft in den Jahren von 1879 bis inclusive 1892.* Stern & Steiner, 1895.

Zerlik, Alfred. "P. Joseph Walcher, S.J." *Mühlviertler Heimatblätter* 9/10 (1964): 138–41.

Ziegler, Anton, ed. *Gallerie aus der Österreichischen Vaterlandsgeschichte in Bildlicher Darstellung.* Vienna, 1837.

INDEX

Note: Page numbers in *italics* indicate figures.

Advisory Commission for Protective Measures Against Flood Dangers, 142–43
alkali soils, 148
Andrásfalvy, Bertalan, 8–9, 140
Andrews, John, 36, 38
Ausgleich. *See* Compromise (Ausgleich) of 1867
The Austro-Hungarian Monarchy in Word and Image, 181–82
Az Arany Ember (Jokai), 29

Baross, Gábor, 77, 107–8, 177
baths/bathing, 159–60, 172–73, 179–80. *See also* swimming/swimming facilities
Bavaria, 39, 63, 92, 93, 102, 137, 138; steamboat company 105, 168–69, 176
beautification club, 161
Beszédes, József, 33

birds, 72–73
Black Sea, 28, 39, 62, 63, 96, 210n131, 220n65
bridges, 25, 42, 46, 47, 50–56, 63–64, 73, 98, 120, 123, 149, 150, 153, 161, 166, 170–73, 182, 195–99; Chain Bridge, 149, 195n99; Elisabeth Bridge, 50, 54, 55, 196n113; Margaret Bridge, 161; pontoon, 25, 54, 160; stone bridges, 52; suspension, 52
Bruck, Karl Ludwig von, 102, 103, 133
Budapest, 55, 149–51; fairs, 115; Franz Joseph Bridge, 52, 53; grand renewal works, 153–54; Millennial Exhibition, 52, 54, 115; promenades and quays, 52, 161; recreation, 160–63; tourism, 115; trade and traffic statistics, 105–6

255

Index

Budapest Metropolitan Board of Public Works, 153, 157, 218n20
Budapesti Hírlap, 70, 104
Building Committee (Építési Bizottság), 156–57

canals, 31–35, 77–85; Canal Law of 1901, 84; Danube-Moldau-Elbe Canal, 80; Danube-Oder Canal, 78, 80, 86, 87; Danube-Sava Canal, 78; Danube-Tisza Canal, 83; drainage, 60, 83, 126, 137, 139; Franz-Joseph Canal, 78; irrigation, 71–72, 83, 139; legislation, 78, 81–82; Main-Rhine-Danube Canal, 87; network, 80; route preferences, 82; Sárvízi Csatorna Társulat, 193n37; Weichsel-Dneistr-San-Canal, 84
ceremonies, 42, 46–47, 49, 50, 52, 54, 56
Charles VI, 26, 28
cholera, 104, 132, 140, 156, 219n35
cities on Danube: Melk, 114, 129, 137–8, 143, 156, 158, 160; small towns, 92, 111–13, 116; urban growth as trade driver, 110–12. *See also* Budapest; Győr, Hungary; Linz, Austria; Vienna
civic engagement, 11, 13, 19, 20, 42, 74, 76, 116, 163, 181–83, 186; Capital Public Works Advisory Council, 50; flood prevention, 20, 148, 170–1
civil society, 59, 86, 128, 133; Flood Committee, 143–44; Private Humanities Association, 159; Viennese Association for Friends of Natural Sciences, 66

climate change, 147, 187–88
coal, 14–15, 97, 100–1
Compromise (Ausgleich) of 1867, 37, 50, 73–74, 80, 102, 104, 107, 118, 135, 170, 189n5, 222n6
Crimean War, 49, 71
customs union, 28, 74, 80, 94, 100, 102, 106–7
Czoernig, Carl Freiherr von, 7, 69, 71, 134

Danube, 16–17, 184–86; as integrative force, 19, 21, 58, 81, 84, 86, 88, 101–10, 112, 131, 186; map of, 57; regulation (*see* regulation); revitalization, 119; rewilding, 147; trade routes *see* trade; traditional transport on, 91–94; transforming, 10–11; tributaries, 11, 33, 58, 76–77, 86; usage, 8–10, 16, 119, 159; wartime strategic importance, 85–86
Danube Association (Donauverein), 3–5, 10–12, 76, 81, 137, 173–74, 176; publications, 13
Danube Canal, 52, 87, 155, 158, 160, *162*, 163
Danube City (Donaustadt), 152–53
Danube Conference, 85–86
"Danube Empire" concept, 6–8
Danube Navigation Act, 104
Danube Regulation Commission (DRC), 50, 55, 133, 134, 138–39, 172, 221n88; Donau-Regulirungs-Commission, 50
Danube Steam Navigation Company (DDSG), 36–41, 62–63, 64, 74–75, 96–110, 112–17, 150, 161, 167, 169, 172; annual busi-

Index

ness report of 1867, 222n6; coal consumer, 97, 112; commercial dominance and monopoly, 38, 96–97, 104–5; establishment, 37; fleet expansion, 97–98; *Franz I* steamboat, 166; freight traffic, 109–10; Hungarian competition and, 104–5, 109, 185; Hungarian nationalist against, 107–8; imperial authorities and, 38–39, 113–17; infrastructure building, 152; local community petitions regarding, 112–13, 115–16; *Maria Anna* steamboat, 38; Óbuda shipyard, 152; passengers safety and convenience, 115–16; statistics, 98, 102, 107, 110; subsidy negotiations, 108, 117; tourism and world's fairs role, 114–15; traffic, 168, 176, 195n98; tugboat, 97

Danubian Confederation, 7

diseases, 155; cholera, 104, 132, 140, 156, 219n35; malaria, 27, 156; typhoid, 158–59. *See also* health

Donaustadt, 152–53, 182, 217n4. *See also* Vienna

Donauverein. *See* Danube Association (Donauverein)

drainage, 27, 34, 60, 63, 124, 126, 135, 150; canals, 60, 83, 126, 137, 139; companies, 10, 133, 141; ecological consequences, 10, 71–72, 87, 140–41; landowners and, 71–72, 139, 140; peasants and, 9–10, 33–34, 61, 140, 148; Tisza valley, 10, 27, 34, 71–72, 139

DRC. *See* Danube Regulation Commission

effluents. *See* sewage systems

Elisabeth (Empress), 44, 46, 47–50, 52, 54–56, 69, 144–45; assassination, 55; devotion to, 55–56

embankments, 46, 49, 64–65, 69, 70, 133–48; causing floods, 9, 66, 136; failure and destruction of, 125, 130–31; flood prevention/protection, 6, 12, 123–26, 170–71; labor programs for building, 132–33; volunteer organizations maintaining, 142–43

energy, 14, 16–17, 33–34, 58–59, 94, 100, 112, 118, 186–88. *See also* coal; firewood

engineering directorates, 32, 33, 64, 66, 68, 77, 123, 133, 142

eudaemonic principle, 56, 67

European Union, 87, 147

Fabrikarm (factory arm), 164, 165, 166, 169, 171–75

factories, 10, 35, 50, 52, 108, 112, 136, 144, 149, 150, 152, 154, 165–66, 173, 177, 182

Ferdinand (Crown Prince), 13, 36, 38; (Emperor), 38, 39, 44, 62, 99, 101, 126, 158

firewood, 23, 100, 111

fishing associations, 108–9

fishing communities, 9, 10, 29, 61, 140–41, 148, 185

floodplains, 146–47

floods/flooding, 19–20, 67–68, 148; cholera outbreak, 156; climate factors causing, 121–23, 156,

257

Index

170; early warning systems, 129–31; economic and political rationale control of, 137–39; engineering solutions increasing risks, 122, 136; impact on agricultural communities, 140–41; imperial family and, 42–46; inadequate management of, 123–26; institutional development for response to, 142–46; Maros River, 42–43; modernity of prevention, 146–47; preventative measures, 44, 146–48; reforms and, 120–23, 131–41; relief, 127–28, 144–46; rescue operations, 143–44; ship mills and, 9; threat, 131–32, 146; transnational cooperation, 147; transnational scientific networks, 127–30

fossil fuels, 15, 97, 100–101, 118. *See also* coal

Franz Canal, 33, 34, 78. *See also* canals

Franz (emperor), 25, 34–35, 63, 67, 125, 159

Franz I steamboat, 23–25, 36, 166

Franz Joseph, 24, 34, 39, 40, 44, 47–55, 67, 80, 83, 86, 115, 120, 123, 142, 158, 161; blasting of *Wirbel* and *Strudel*, 68–69; and early warning systems, 129–30; flood relief and response, 44–46, 144–45; ice skating, 163; motto, 4, 21; National Diet, 83; public openings and commemorations, 52–55; role in centralization of hydraulic engineering, 68–69, 71, 133;

Suez Canal, 77–78; uprisings of 1848–49, 49, 67, 133. *See also* Compromise (Ausgleich) of 1867

Franz Joseph Bridge, 52–54, 195–96n99; postcards, 54; *Vasárnapi Ujság* on, 53, 54

Frederick II of Prussia, 26, 27

free trade, 27–28, 101

freight traffic/transport, 39, 40–41, 63, 87–88, 98–100, 102–17; traditional transport on Danube, 91–94

French Revolution, 25, 67

Fritsch, Anton, 72

greenhouse gases, 187. *See also* climate change

Győr, Hungary, 20, 166–81; flood/flooding, 170–71; grain export market, 166; historical map, 165; recreation, 180–81; steam navigation, 167

Győr Steam Navigation Company, 169, 178, *178*

Habsburg Empire, 3, 8, 26, 119; as a connected community, 118–19; "Danube Empire" concept, 6–8; energy transition, 118; engineering works as symbol of unity, 57–59, 73–74, 89–90, 176, 181–82; food supply, 6; imperial displays, 47–49, 50–54; industrialization, 14–16; infrastructure, 51–56; population boom, 6; reforms, 56, 66–68; riparian communities, 14; scholars, 16; uprisings of 1848–49, 49. *See*

258

Index

also floods/flooding; Franz Joseph; regulation; steamboats; trade; transportation
health, 154–59; industrialization and, 154–55. *See also* diseases; sewage systems
Hieronymi, Károly, 76
Hunfalvy, János, 5
Hungarian Royal River and Sea Joint-Stock Company (MFTR), 41, 108, 178, 194n70

ice dams, 12, 66, 122, 156
Imperial-Royal Trade Ministry, 40, 67, 70, 74, 116, 131
industrialization, 14–16, 151–54. *See also* factories; trade
Industrial Revolution, 14, 118; Second Industrial Revolution, 15. *See also* industrialization
Inland Waterway Congress, 79
Institutum Geometrico-Hydrotechnicum, 32
irrigation canals, 71–72, 83, 139

Jägerzeile, Vienna, 161
Jászi, Oszkár, 7
Johann, Archduke, 44
Jokai, Mór, 29, 149
Joseph, Archduke, 24, 36, 37, 42–43, 62, 65, 66, 156, 193n37
Joseph II, 6, 32, 49, 94, 161

Kiss, József, 33, 34, 60–61
Klun, Vinzenz, 57–60, 72
Koerber, Ernest von, 80–81
Kossuth, Lajos, 7, 63

Linz, Austria, 20, 163–76; flood/flooding, 165, 171; freight traffic, 175, 176; passenger traffic, 175, 176; recreation, 179–80; transshipment/transit hub, 174–76; warehouse, 176
Little Ice Age, 12, 66, 91, 121, 124, 156
Lower Austria 10, 93, 110, 113, 116, 139, 143, 145, 159; business and commercial groups, 75, 153, 159; government, 76, 81, 101, 114, 134, 137–138, 142, 172; trade and traffic, 113–114; 145

Maria Theresa, 6, 9, 26–35, 78, 94, 155
Millennial Exhibition of Hungary, 52, 54, 115
Mohammed Ali, 78

Napoleon III, 47
natural disasters, 19–20. *See also* floods/flooding
Navigation Directorate, 28–29, 30, 32, 60
Newcomen, Thomas, 14
Nicholas I of Russia, 49
Nicholas II of Russia, 47
nongovernmental organizations, 75

Óbuda, 37, 152. *See also* Danube Steam Navigation Company (DDSG)
Oelwein, Arthur, 80, 173
On Danube Navigation (Széchényi), 37
Ordinarischiffe, 91
Ottoman Empire, 17, 26–27, 220n65

Index

Palacký, František, 89–90
paper industries, 100. *See also* factories
Pasetti, Florian, 70, 71
peasants, 9–10, 61, 140, 148; access to arable land, 34; revolts/uprisings, 33–34, 132. *See also* floods/flooding
Pester Lloyd, 52–53
Pesti Hírlap, 167, 180
Pesti Napló, 52, 73, 133–34
Peter the Great, 31–32
Prater, 52, 161
Prussia 26–7, 73, 79, 80
public celebrations: bridge openings, 50–54; Elisabeth's 1854 steamboat arrival, 47–49; naming infrastructure after Habsburgs, 51–52, 54–55

quays, 42, 46, 50, 52, 133, 161; promenades, 161; workforces, 150

Rába River, 75, 137, 157, 170–71, 175, 180–81
railways/rail networks, 35, 39, 40, 72, 76, 79, 82, 84, 153–54; as competition for river transport, 103–8, 112–14, 116, 118, 167–70; Oelwein on, 80; popularity and importance, 97; Railway Concession Law, 103–4. *See also* trade; waterways
rapids, 29–31, 68–69
reclamation, 32, 124, 134, 139
recreation, 159–63, 179–81; ice-skating, 163, 179; promenade, 161. *See also* swimming/swimming facilities

regulation, 3–6, 11–14, 19, 32–33, 38; advocates for, 74–76; centralization of, 66–68; civic engagement, 181–83, 186; Czoernig on, 7; environmental consequences, 10, 185; fishing and, 10, 29; funding increases, 76–77; Law VIII, 76; modern parallels, 86–88; resistance to, 9–10; as state policy, 27–31, surveys, 33; as symbol of imperial unity, 57–59, 73–74, 89–90, 176, 181–82; urban, 50–52. *See also* canals; drainage; embankments; floods/flooding; waterways
Renan, Ernest, 20, 223n9
rivers, 16–18, 57. *See also* Danube; floods/flooding; regulation; *specific river*
River Supervisory Offices, 68
Romania, 40–41, 85, 114, 147, 150
rowing, 162–63; Duna Rowing Club, 163; Győr Boat Club, 180; National Boat Club, 163
Royal Ministry of Public Works and Transport (KMKM), 5, 73, 74, 76, 104–5, 107, 112, 132, 198n46, 200n84
Rudolf (Crown Prince), 6, 13, 44, 72, 181
Russia, 31–32, 39, 40, 47, 49, 62, 63, 79, 90, 133, 220n65; waterways, 79

salt: commodity/trade, 23, 35, 91, 92, 102, 112, 118; tax revenue, 62–63
sewage systems, 6, 156–57, 170, 174; effluents, 10, 155, 158;

Index

Water Law of 1855, 158. *See also* waste disposal/dumping
shipbuilding industry, 17, 29–30, 100, 105, 169. *See also* steamboats
ship mills, 17, 122, 105, 171; navigation and, 9–10, 29–30, 105
skating. *See* recreation: ice-skating
Smith, Adam, 27
solidarity, 19, 127, 144, 147, 181
steamboats, 11, 13, 19, 38, 40, 96–100, 103–14, 116–18, 150, 152, 153, 161, 166–70; companies, 40, 96, 105, 109, 118, 174, 176, 178–79; disrupting habitats, 10; on American rivers, 36; *Franz I*, 23–25, 36, 166; *Franz Joseph*, 47–49; stations, 52, 73; trials, 61. *See also* Danube Steam Navigation Company (DDSG)
steam engines, 14–15, 23
steam navigation, 36–41, 61–63, 95–96, 166–67, 185; business connections, 152; early years, 23–25, 99–100
Strudel rapids, 30–31, 68–69. *See also* rapids
Suess, Eduard, 3–4, 76
swimming/swimming facilities, 143–44, 159, 160, 166, 172, 179–81
Széchenyi, István, 37, 62–63, 65–66, 71, 94, 95, 97, 127, 162, 166, 167

telegraph, 116, 129, 131
tourism, 88, 98, 114–15, 117
trade, 3, 5, 7, 9, 36–41, 50, 55, 73–75, 90, 94–99, 116–18, 152–54, 166–70; commercial infrastructure, 6; customs union, 28, 74, 80, 94, 100, 102, 106–7; free, 27–28, 101; grain, 29, 103–4, 166–67, 170; integration through, 92–93, 102–3, 118; international/global, 61, 62, 118; joint framework for, 74; obstacles, 27–28, 107–9; preferential arrangement, 28; restrictions, 93–94; Revolutions of 1848 impact on 101–2; salt, 35; statistics, 98, 102, 105, 107, 110, 117; Theresa and, 26–29; traditional, 91–92; transimperial, 92, 109, 174; urban growth as driver, 110–12; Vienna as hub, 93, 110–11. *See also* waterways
transnational cooperation, 19, 147, 151, 188
transportation, 3, 10, 18, 35, 37, 41, 46, 55; freight, 39, 40–41, 63, 87–88, 98–100, 102–17. *See also* railways/rail networks; waterways
Tulla, Johann Gottfried, 68

Upper Austria 48, 82–83, 92, 163, 165; government, 77, 81, 83, 106, 113, 115, 116, 138, 173–74; flood protection/relief, 123, 129, 131, 145
urban: growth, 110; transformation and development, 149–54. *See specific cities*
US Army Corps of Engineers, 33

Vásárhelyi, Pál, 33, 161
Vedres, István, 34

261

Index

Vienna, 3–4, 44, 93, 96–98, 102, 110–11; development, 153, 182; Donaustadt, 152–53, 182, 217n4; and early warning systems, 129–31; flood protection plans of, 132–36, 142–44; *Franz I* steamboat arrival in, 23–25; housing crisis, 155; migration to, 150–51; new infrastructure, 51–52; recreation, 160–61, 163; sewage and health, 155–57; Trieste road building, 28; water supply, 158–59; World's Fair of 1873, 52, 115, 158

Vienna River, 50, 156

Voltaire, 31

Walcher, Joseph, 30, 33

warehouses, 153–54, 176–78. *See also* trade

waste disposal/dumping, 155, 159. *See also* sewage systems

water brigades, 143, 144

water levels, 12, 88, 94, 128–29, 130, 131, 136

water supply system, 158–59

waterways, 11–13, 19, 34, 61, 63, 70–92, 95, 105–9, 139, 163, 176; American, 27–28; crisscrossed, 46; drainage projects, 135; dumping of waste, 158; expansion, 36, 71, 80; German military rationale, 85; Oelwein on, 80; polluting, 108; shipping costs, 79. *See also* canals; railways/rail networks

whirlpools, 30–31, 91. *See also* rapids

Wiener Neustädter Canal, 34–35, 78, 163. *See also* canals

Wilhelm II of Germany, 47

Wirbel rapids, 30–31, 68–69. *See also* rapids

World's Fair of 1873, 52, 115, 158

World War I, 85–86, 138, 159, 160, 163, 177, 184

Youth Games and Physical Fitness, 179–80

Zentralanstalt für Meteorologie und Geodynamik (Central Institute for Meteology and Earth Magnetism), 130

www.ingramcontent.com/pod-product-compliance
Lightning Source LLC
Chambersburg PA
CBHW031411290426
44110CB00011B/340